Electroanalysis

Christopher M.A. Brett
Ana Maria Oliveira Brett

Department of Chemistry, University of Coimbra, Portugal

Series sponsor: **ZENECA**

ZENECA is a major international company active in four main areas of business: Pharmaceuticals, Agrochemicals and Seeds, Specialty Chemicals, and Biological Products.

ZENECA's skill and innovative ideas in organic chemistry and bioscience create products and services which improve the world's health, nutrition, environment, and quality of life. ZENECA is committed to the support of education in chemistry and chemical engineering.

ÆDES CHRISTI
in Academia Oxoniensi

OXFORD   NEW YORK   TOKYO
OXFORD UNIVERSITY PRESS
1998

Oxford University Press, Great Clarendon Street, Oxford OX2 6DP

Oxford   New York
Athens   Auckland   Bangkok   Bogota   Buenos Aires   Calcutta
Cape Town   Chennai   Dar es Salaam   Delhi   Florence   Hong Kong   Istanbul
Karachi   Kuala Lumpur   Madrid   Melbourne   Mexico City   Mumbai
Nairobi   Paris   São Paolo   Singapore   Taipei   Tokyo   Toronto   Warsaw
and associated companies in
Berlin   Ibadan

Oxford is a trade mark of Oxford University Press

Published in the United States
by Oxford University Press Inc., New York

A catalogue record for this book is available from the British Library

Library of Congress Cataloging in Publication Data
Brett, Christopher M. A.
Electroanalysis / Christopher M.A. Brett, Ana Maria Oliveira Brett.
(Oxford chemistry primers; 63)
Includes index.
1. Electrochemical analysis.   2. Electrochemistry.   I. Brett, Ana
Maria Oliveira.   II. Title.   III. Series.
QD115.B658 1998   543'.0871–dc21   98-22282
ISBN 0 19 854816 8 (Pbk)

Typeset by EXPO Holdings, Malaysia

Printed in Great Britain by
The Bath Press Ltd, Bath

Series Editor's Foreword

Oxford Chemistry Primers are designed to provide clear and concise introductions to a wide range of topics that may be encountered by chemistry students as they progress from the freshman stage through to graduation. The Physical Chemistry series will contain books easily recognised as relating to established fundamental core material that all chemists will need to know, as well as books reflecting new directions and research trends in the subject, thereby anticipating (and perhaps encouraging) the evolution of modern undergraduate courses.

In this Physical Chemistry Primer, Ana and Christopher Brett present a stimulating and self-contained introduction to the principles and applications of *Electroanalysis*. The Primer will interest all students (and their mentors) who require an easy-to-read yet authoritative introduction to this increasingly important topic.

Richard G. Compton
Physical and Theoretical Chemistry Laboratory, University of Oxford

Preface

In recent years, many parts of chemistry have undergone significant changes with the advent of computer-controlled instrumentation, automation, miniaturization and novel fabrication procedures. This has had a large impact on quantitative measurements and on the speed of data acquisition and analysis.

The objective of this primer is to present modern electroanalysis, the application of electrochemistry to analytical problems, at an advanced under-graduate level. The book has been written as a self-contained text which describes the basic principles of the necessary electrochemistry and then moves on to electrochemical sensors, with an overview of future trends. Many exciting developments have occurred in recent years and which are indicated—applications can be in areas as diverse as industry, health and the environment.

It is our hope that the book will kindle the enthusiasm of the reader to seek more detailed information concerning new developments and strategies and to use electroanalytical techniques.

<table>
<tr><td>*Coimbra*
February 1998</td><td align="right">C.M.A.B.
A.M.O.B.</td></tr>
</table>

Contents

1 Introduction   1

2 Electrochemical principles   8

3 Potentiometric sensors   37

4 Voltammetric sensors   48

5 Applications   69

Appendix 1. Data analysis   76

Appendix 2. Standard electrode potentials   83

Index   87

1 Introduction

1.1 Electroanalysis, its scope and role in analytical chemistry

Electroanalysis can be defined as the application of electrochemistry to solve real-life analytical problems, and the aim of this book is to demonstrate the different potentialities of electrochemical methodologies. Traditionally, analytical chemistry has been associated mainly with the measurement of amounts (concentrations) of elements or species. In recent years there has been a tendency, aided by the development of more sophisticated techniques, to combine identification with quantification and, in some cases, to obtain other time-dependent information on kinetics of degradation, uptake by organisms, etc. simultaneously with quantitative measurements. Each analytical technique has a specific purpose and range of application, and care must be taken to choose and evaluate which is the most appropriate in a given situation, as will be described below.

Electroanalytical measurements offer a number of important potential benefits which may or may not be realisable in given situations:

(a) selectivity and specificity, i.e. probing of speciation, resulting from the applied potential

(b) selectivity resulting from choice of electrode material

(c) high sensitivity and low detection limit resulting from the use of complex applied potential programmes

(d) possibility of giving results in real time, or close to real time, particularly in flow systems for on-line monitoring

(e) application, as miniaturised sensors, in situations where other sensors may not be usable

Naturally, the principal criterion is that the species which it is desired to measure should react directly (or indirectly through coupled reactions) at, or be adsorbed onto, the electrode. Electroanalytical measurements can only be carried out in situations in which the medium between the two electrodes making up the electrical circuit is sufficiently conducting. In general, this means the liquid phase, although there are important exceptions based on solid-state conduction and in supercritical fluids. Nevertheless, for measurements made at zero current, the conductive path between the electrodes can be highly resistive. These constraints are the main limitations to the application of electrochemical sensors, although other more specific ones do arise in particular situations, as will become evident further on in this book.

Thus, electroanalysis is complementary to other forms of analysis. An often made comparison is with the widely used atomic absorption spectrometry

(AAS) in its more modern versions with graphite furnace or inductively coupled plasma (ICP), which show similar detection limits to electrochemical detectors for heavy metals. However, AAS responds to the total amount of a given element; often elaborate digestion procedures are needed to prepare the sample and the measurement has to be done in the laboratory. AAS says nothing about how much exists in what oxidation state or what is the particular species, i.e. the speciation, and whether the species is labile. The electrochemical approach applied to solution samples will give a rapid answer, without digestion, as to the labile fraction of a given element in a particular oxidation state, and the experiment can be performed on-site in the field. In cases when two comparable analytical methods can be applied the type of information and the rapidity of the response will often decide the method of choice.

Other criteria which have to be borne in mind are linked to the cost of performing the analyses and how competent the technical operator needs to be. With present trends in computer control of analytical experiments, the level of expertise necessary in order to obtain results is becoming less. The other side of the coin is that apparently excellent, but wrong, results can be obtained by operators without sufficient background knowledge to be able to interpret correctly their raw results and know the limitations of the procedures.

The purpose of this book is to give the background knowledge, the elements of electrochemical techniques applied in modern-day electroanalysis and some applications in order to show the power of these techniques. This is becoming more and more important as many chemists without a training in electrochemistry or scientists in materials, industrial, biological and medical firms are beginning to use electroanalytical techniques in the laboratory.

1.2   Types of electroanalysis

There are essentially three types of electroanalytical measurement that can be performed:

(a)   conductimetric

(b)   potentiometric

(c)   amperometric and voltammetric

In *conductimetry*, it is the concentration of charge which is obtained through measurement of solution resistance—this is therefore not species selective. It can, however, be useful in situations where it is necessary to ascertain, for example, whether the total ion concentration is below a certain permissible maximum level or for use as an on-line detector after separation of a mixture of ions by ion chromatography.

In *potentiometry*, the equilibrium potential of an indicator electrode is measured against a selected reference electrode using a high impedance voltmeter, i.e. effectively at zero current. Thus, the current path between the two electrodes can be highly resistive. If an inert redox electrode such as platinum is used as indicator electrode then its potential will be a mixed potential, the value of which is a function of all species present in solution and their concentrations. By judicious choice of electrode material, the selectivity of the response to one particular ion can be increased, in some cases with only very minimal interference in the measured potential from other ions. Such electrodes are known as ion-selective electrodes, an example being the pH electrode. Detection limits are of the order of 100 nanomoles per litre of the total concentration of the ion present in a particular oxidation state, although down to 10 picomolar differences in concentration can be measured— effectively this is a measurement of resolution and is dependent on the accuracy and precision of the external measuring circuit.

In *amperometry*, a fixed potential is applied to the electrode, which causes the species to be determined to react and a current to pass. If this potential is conveniently chosen, then the magnitude of the current is directly proportional to concentration. Additionally, if steady-state convection is employed, as in flowing streams, then a constant current is measured if the concentration of electroactive species is uniform. Microelectrodes also permit the attainment of steady-state currents. Detection limits in the micromolar region can be obtained.

More information and lower detection limits can usually be gained from the use of *voltammetry*. Here, the current is registered as a function of applied potential. In this way, several species which react at different applied potentials can be determined almost simultaneously in the same experiment without the need for any previous separation step. Very low detection limits of down to picomolar concentrations can be reached using state-of-the-art instrumentation and preconcentration of the analyte on the electrode surface. A major part of this primer will deal with voltammetric techniques, which have wide application and enable investigation of the electrode process at the same time as quantitative data are being obtained.

1.3 General criteria for design of electrochemical experiments

A number of experimental design factors have to be considered if it is decided to perform an analytical experiment with electrochemical detection. Amongst the most important are:

(a) for amperometric and voltammetric sensors, is the species electroactive within a reasonable potential range? To what extent does the addition of an inert (i.e. supporting) electrolyte (if not already present) to carry the current perturb the equilibria in solution?

(b) for potentiometric sensors, is there an adequate electrode material, free from interferences?

(c) can the concentration be determined with sufficient accuracy and precision?

(d)   are the measurements sufficiently reliable and repeatable?

(e)   is the response time of the sensor sufficiently fast?

(f)   is there drift or diminution of sensor response with time?

(g)   is calibration simple and easy to perform?

(h)   is the detection limit sufficiently low?

The answers to these questions depend on the technique employed, the electrode material and the electrode and cell configuration.

The useful potential ranges of electrode materials are determined by oxidation or reduction of the solvent (solvent decomposition), decomposition of the supporting electrolyte, electrode dissolution or formation of a layer of insulating/semiconducting substance on its surface. Electrode materials for voltammetry must conduct electrons. Thus their choice is limited to metals, other solids with metallic conductivities and good semiconductors. Usually it is also desired that the electrode material is inert in the region of potential in which the electroanalytical determination is carried out. Mercury was the first metal to be used extensively, in the form of the dropping mercury electrode; however, mercury's useful potential range is limited by its oxidation which means that, essentially, only reductions can be investigated. Thus, solid electrode materials were developed which permit oxidation reactions to be studied. These questions will be further discussed in Chapter 2.

The *reliability* and *repeatability* of experiments can be aided by assuring a constant flux of electroactive species to the electrode. This is done by using controlled convective flow over the electrode, known in such situations as a *hydrodynamic electrode*, particularly in flow streams (an example is a chromatography detector), or by creating a sufficiently high concentration gradient, as in *microelectrodes*. The additional advantage of this approach is that, because of the greater mass transport, *sensitivity* is increased and *detection limits* are lowered.

Detection limits can also be affected by other electrode reactions which can occur in the same potential range. The most prevalent of these, and one which interferes with the study of reduction processes, is the reduction of oxygen, since its solubility in solutions open to the atmosphere is up to $10^{-4}M$. This means that oxygen must be removed from the solution in these cases, usually by passage of an inert gas, prepurified nitrogen or argon, through the solution to diminish the oxygen partial pressure to a very low value. There are some modern voltammetric techniques which obviate the need for oxygen removal, but it is a general problem which must be considered.

The poisoning of electrode surfaces has been one of the main limitations to the widesperead use of electroanalysis by non-experts. Apart from regeneration of the surface through cleaning or polishing, there are three main ways of reducing these problems which are all being pursued at present. First, use of a suitable technique such that the time during which adsorption of the poisoning substance occurs is reduced to a minimum via reduction of the contact time of the analyte with the electrode or the time during which the potential corresponding to adsorption is applied. Secondly, making the surface of the electrode exposed to the solution incompatible with adsorption by modification or covering with a specially designed membrane. Thirdly, use of disposable electrodes which are only employed for a short period of time during which adsorption problems are negligible.

1.4 How to choose the electroanalytical technique

A choice of electroanalytical technique only really becomes necessary when it is decided to use voltammetry and arises due to the availability of instruments which offer a wide range of techniques, related to the different types of applied potential waveform which are possible. They can be divided into three classes:

(a) linear sweep methods

(b) step and pulse techniques

(c) ac voltammetric techniques (based on electrochemical impedance) together with combinations of these three classes.

Modern electrochemical instrumentation tends to be microprocessor-controlled and so applied waveforms are, in general, digitally based without digital to analogue conversion. This has been particularly used in the application of step and pulse techniques. Linear sweep is nowadays often implemented as a step technique, consisting of a succession of small potential steps in a staircase rather than a true ramp, which can lead to erroneous results if due care is not taken. Ac voltammetry has been less used but gives equivalent information to step and pulse techniques.

The time-scale of the perturbation applied to the electrode can be varied considerably within each technique, which may or may not allow the attainment of equilibrium at the electrode surface. Although usually this equilibrium will lead to a larger current response, this is at the expense of any kinetic information.

These considerations and others must be taken into account when designing an appropriate procedure for an electroanalytical experiment.

1.5 Analysis of electroanalytical data

A correct interpretation and analysis of data obtained in electroanalytical experiments is obviously crucial. It is also important to repeat measurements in order to improve precision and ensure accuracy by periodic calibration. When many factors can influence the experimental response, chemometrics can aid in deconvoluting the response. One of the most important aspects, particularly nowadays due to the increasing preoccupation with environmental levels of many species, is the detection limit. This is defined with respect to the noise level of the signals obtained, being equal to three times the standard deviation of the slope of the calibration plot (see Appendix 1). In this way, impressive detection limits can often be obtained, since this criterion is, in essence, a measure of the resolution of signals. In many real cases, it is of limited use owing to the fact that there are other indirectly contributing factors to the general signal profile which may vary with time, leading to a practical detection limit (i.e. down to what level can the signals for different concentrations be statistically distinguished) which may be one or even two orders of magnitude higher.

Bibliography

Some electroanalytical chemistry texts and electrochemistry texts which contain sections with emphasis on electroanalysis are given below in chronological order.

R.N. Adams, *Electrochemistry at solid electrodes*, Dekker, New York, 1969.

A.J. Bard and L.R. Faulkner, *Electrochemical methods, fundamentals and applications*, Wiley, New York, 1980.

A.M. Bond, *Modern polarographic methods in analytical chemistry*, Dekker, New York, 1980.

B.H. Vassos and G.W. Ewing, *Electroanalytical chemistry*, Wiley, New York, 1983.

J. Koryta, *Ions, electrodes and membranes*, Wiley, Chichester, 1991.

M.R. Smyth and J. Vos (eds), *Analytical voltammetry*, Elsevier, Amsterdam, 1992.

C.M.A. Brett and A.M. Oliveira Brett, *Electrochemistry. Principles, methods, and applications*, Oxford University Press, Oxford, 1993.

J. Wang, *Analytical electrochemistry*, VCH, New York, 1994.

D.T. Sawyer, A. Sobkowiak and J.L. Roberts, *Experimental electrochemistry for chemists*, 2nd edn, Wiley, New York, 1995.

P.T. Kissinger and W.R. Heineman (eds), *Laboratory techniques in electroanalytical chemistry*, 2nd edn, Dekker, New York, 1996.

Texts treating more specific aspects of electroanalysis are:

F. Vydra, K. Stulik and E. Julakova, *Electrochemical stripping analysis*, Ellis Horwood, Chichester, 1976.

J. Wang, *Stripping analysis: principles, instrumentation and applications*, VCH, Deerfield Beach, Florida, 1985.

R. Kalvoda (ed.), *Electroanalytical methods in chemical and environmental analysis*, Plenum, New York, 1987.

K. Stulik and V. Pacakova, *Electroanalytical measurements in flowing liquids*, Ellis Horwood, Chichester, 1987.

A.P.F. Turner, I. Karube, and G.S. Wilson (eds), *Biosensors: fundamentals and applications*, Oxford University Press, Oxford, 1987.

J. Wang, *Electroanalytical techniques in clinical chemistry and laboratory medicine*, VCH, New York, 1988.

A.E.G. Cass (ed.), *Biosensors: a practical approach*, IRL Press, Oxford, 1990.

J.P. Hart, *Electroanalysis of biologically important compounds*, Ellis Horwood, Chichester, 1990.

T.R. Yu and G.L. Ji, *Electrochemical methods in soil and water research*, Elsevier, Amsterdam, 1993.

Kh. Brainina and E. Neyman, *Electroanalytical stripping methods*, Wiley, New York, 1993.

2 Electrochemical principles

2.1 Introduction

Electrochemistry is the study of phenomena caused by charge separation. In the context of electrode processes, which are heterogeneous in nature, it deals with the study of charge transfer processes at the electrode/solution interface, either in equilibrium at the interface, or under partial or total kinetic control.

Most of the charge transfer processes are transfer of electrons, which can be represented in the simplest case of oxidised species, O, and reduced species, R, both soluble in solution, by

$$O + ne^- \rightleftharpoons R$$

where O receives n electrons in order to be transformed into R. The electrons in the electrode (a conductor) have a maximum energy which is distributed around the Fermi level: it is only around this level of energy that electrons can be supplied or received. The Fermi level, E_F, can be externally influenced by application of a voltage which either injects electrons into or removes electrons from the conductor.

The process of electron transfer can be physically visualised as follows. For a reduction, the electrons in the electrode must have a minimum energy in order to be transferred from the electrode to the receptor orbital in O; for oxidation the energy of the electrons in the donor orbital of R must be equal to or higher than the electrode's Fermi level in order to be transferred to the electrode (see Fig. 2.1). These electron energies correspond to the electrode potentials which can be measured against a suitable reference. The corresponding energies of the orbitals in the O/R pairs are their redox potentials, which are not necessarily exactly the same, due to small energy differences which can appear over the distance from the electrode surface to the species O or R.

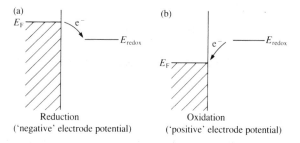

Fig. 2.1 Electron transfer at an inert metal electrode. E_F is altered by the applied potential, thus facilitating (a) reduction or (b) oxidation.

It can be seen from the above, that the current, which is directly proportional to the number of electrons transferred, is therefore proportional to the concentration of the electroactive species close to the electrode.

2.2 Electrochemical cells

In a real situation, both reduction and oxidation must occur to an equal extent, so as not to violate the law of conservation of energy. Thus experimental assemblies consist of a container of variable geometry, the electrochemical cell, in which two independent electrodes are immersed in a conducting phase, usually a solution, and are linked through the solution and through external wires, electrical loads or batteries so as to form an electrical circuit. Oxidation or reduction reactions, known as electrochemical reactions or half-reactions, occur at each electrode, consuming electrons at one electrode and supplying them at the other so that the overall chemical reaction does not involve any net consumption of electrons.

The current flowing through the electrical circuit can result either from the occurrence of spontaneous electrode reactions (*galvanic cell*), or from the supply of electrical energy in the external circuit which causes reactions to occur at the electrodes (*electrolytic cell*) via changing the electron energy in the electrodes, as indicated in the previous section.

Electrode potentials

The Nernst equation describes the variation of electrode potential for each electrochemical reaction with the activities of oxidised and reduced species, a_{O_i} and a_{R_i}, under conditions of equilibrium at the electrode surface, and is given by

$$E_{eq} = E^{\ominus} + \frac{RT}{nF} \ln \frac{\Pi a_{O_i}^{v_i}}{\Pi a_{R_i}^{v_i}} \tag{2.1}$$

where $E^{\ominus}$ is the standard electrode potential of the electrochemical reaction when all species have unit activity. These reactions are always written as reductions, by convention.

According to accepted practice, the reference potential for standard electrode potentials, $E^{\ominus}$, is the standard hydrogen electrode, the half-reaction for which is

$$H^+ + e^- \rightleftharpoons \tfrac{1}{2}H_2$$

This electrode will be discussed further below. A table of standard electrode potentials is shown in Appendix 2.

Formal potentials

In electroanalytical experiments, the most important quantity is usually the concentration of an electroactive species rather than its activity. This is because the number of electrons transferred, i.e. the current, is directly proportional to the concentration, not to the activity. Thus a form of the Nernst equation in which concentrations appear is often used, which is

$$E_{eq} = E^{\circ\prime} + \frac{RT}{nF} \ln \frac{\Pi[O_i]^{v_i}}{\Pi[R_i]^{v_i}} \tag{2.2}$$

where $E^{\circ\prime}$ is called the *formal potential*. Comparing Eqns 2.1 and 2.2 it can be seen that the formal potential is defined as

$$E^{\circ\prime} = E^{\circ} + \frac{RT}{nF} \ln \frac{\Pi\gamma_{c,O_i}^{v_i}}{\Pi\gamma_{c,R_i}^{v_i}} \tag{2.3}$$

recognising that $a = \gamma c$, where γ is the activity coefficient and c the concentration. In actual fact, E° can be obtained by a series of measurements at different concentrations, extrapolating to zero concentration where the activity coefficient is equal to unity. The occurrence of secondary reactions, such as complexation, can influence the values of $E^{\circ\prime}$. It can be seen from Eqn 2.3 that the formal potential varies according to the electrolyte solution in which the measurements are performed, although often this difference is only of the order of a few millivolts and is thus not very significant.

Cell potential

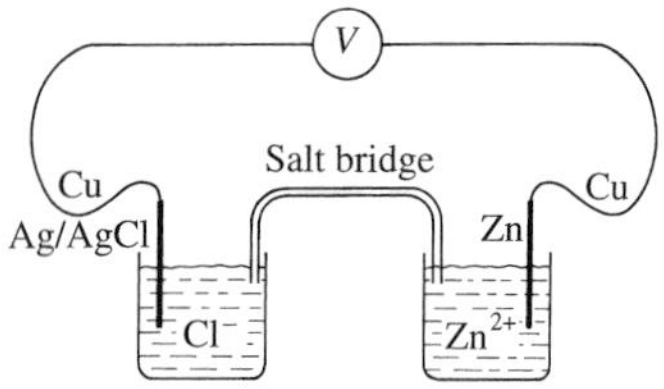

A typical example of an electrochemical cell is

$$\text{Ag} \mid \text{AgCl} \mid \text{Cl}^-(\text{aq}) \, \| \, \text{Zn}^{2+}(\text{aq}) \mid \text{Zn}$$

and is shown in Fig. 2.2. According to IUPAC recommendations, oxidation is assumed to occur in the left half-cell and reduction in the right half-cell. The reduction reactions are:

Left:	$\text{AgCl} + \text{e}^- \rightleftharpoons \text{Ag} + \text{Cl}^-$	$E^{\circ} = 0.222 \text{ V}$
Right:	$\text{Zn}^{2+} + 2\text{e}^- \rightleftharpoons \text{Zn}$	$E^{\circ} = 0.760 \text{ V}$

where E° are the standard electrode potentials. In the left half-cell, electrons produced by oxidation of Ag to AgCl flow through the external Cu wires and voltmeter to the right half-cell where they reduce Zn^{2+} to Zn. The salt bridge, represented by $\|$ in the cell diagram, and which is often a solution of KCl, completes the electrical circuit and serves to maintain electroneutrality in each half-cell.

The expression for the cell potential, E_{cell}, combining the two half-cells, is

$$E_{cell} = E_{right} - E_{left} \tag{2.4}$$

The difference in the energies of the electrons in the two electrodes is being calculated, hence there is a negative sign in Eqn 2.4 and there is no

Fig. 2.2   A simple electrochemical cell.

dependence on the number of electrons transferred in the half-reactions as written. The negative sign is associated with E_{left} since this reaction has to be inverted to become an oxidation. A value of E_{cell} of 0.538 V is calculated if all species are at unit activity. The maximum work that can be obtained is

$$\Delta G = -nFE_{cell} \qquad (2.5)$$

Reference electrodes

Reference electrodes are necessary to provide a stable, drift-free, accurate value of potential as a reference voltage. In cell diagrams, reference electrode couples appear on the left.

Standard electrode potentials, which are written as reductions with respect to the standard hydrogen electrode, correspond to cells of the type

$$H_2(p = 1 \text{ atm}), \text{ Pt} \mid HCl\,(a = 1.0) \parallel Zn^{2+}(aq) \mid Zn$$

where E_{left} is, by definition, zero.

The standard hydrogen electrode is the reference electrode for standard electrode potentials. It consists of a platinized platinum electrode immersed in a solution of hydrochloric acid $(a = 1)$ over which hydrogen gas bubbles at 1 atm pressure. Although it is extremely accurate, it is obvious that it is not convenient to use on a day-to-day basis: apart from practical difficulties there are the dangers associated with hydrogen gas and ensuring that the platinum electrode is sufficiently active. Thus substitutes have been developed, mainly based on metals in intimate contact with a sparingly soluble salt of the corresponding cation. A number of important reference electrode half-reactions are shown in Table 2.1 and typical designs for the widely used calomel and Ag|AgCl electrodes in Fig. 2.3. Note that in all cases the activity of all species (metal, sparingly soluble salt) except one (ion in solution) is essentially unity. This is an important aid in ensuring that the potential of the reference electrode is stable and not susceptible to oscillation.

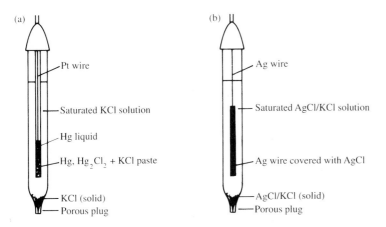

Fig. 2.3 Typical designs for (a) calomel and (b) Ag|AgCl reference electrodes.

Table 2.1   Reference electrode electrochemical reactions in water solvent

Electrochemical reaction	E° / V
$AgBr + e^{-} \rightleftharpoons Ag + Br^{-}$	0.071
$AgCl + e^{-} \rightleftharpoons Ag + Cl^{-}$	0.222
$Hg_2Cl_2 + 2e^{-} \rightleftharpoons 2Hg + 2Cl^{-}$	0.268
$HgO + H_2O + 2e^{-} \rightleftharpoons Hg + 2OH^{-}$	0.098
$Hg_2SO_4 + 2e^{-} \rightleftharpoons 2Hg + SO_4^{2-}$	0.613

Working or indicator electrodes

The working electrode is where the electrode reaction under study is taking place. Electrode materials for use under non-equilibrium conditions must be good electron conductors. Thus one is limited to metals, other solids with metallic conductivity and good semiconductors. Usually it is also necessary that the electrode material is inert in the region of potential in which the electroanalytical determination is carried out.

Table 2.2 shows the approximate potential ranges of three commonly used electrode materials, platinum, mercury and carbon, in common solvents and

Table 2.2   Approximate potential ranges for platinum, mercury and carbon in aqueous and non-aqueous electrolytes

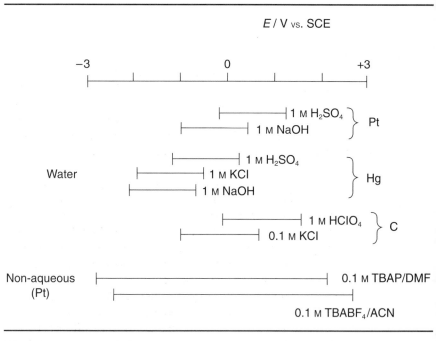

TBAP	tetrabutylammonium perchlorate
DMF	dimethylformamide
TBABF$_4$	tetrabutylammonium tetrafluoroborate
ACN	acetonitrile

electrolytes. The limits are determined by one or more of solvent decomposition (i.e. oxidation or reduction of the solvent), decomposition of the supporting electrolyte, electrode dissolution or formation of a layer of insulating/semiconducting substance on its surface. Changing the solvent or using a mixed solvent can sometimes extend the potential range. Note the comparatively very negative potential limit of mercury in aqueous media due to the high overpotential for the evolution of hydrogen.

Mercury was the first metal to be used extensively for electroanalytical purposes in the form of the dropping mercury electrode. Its cyclic operation—continual drop growth, release and renewal—avoids many of the problems of electrode poisoning in complex matrices. However, oxidation of the mercury itself means that, in practice, only reduction reactions can be investigated (see potential range in Table 2.2), the detection limit can be too high and there are some experimental manipulation difficulties. Thus, solid electrode materials were developed, many of which also permit oxidation reactions to be studied. However, there is no electrode material as good as mercury for studying reductions owing to its extended negative limit—it is often used nowadays as a static drop or as a thin film on a suitable substrate. A type of solid material which has become very common is carbon, in the form of glassy carbon, graphite, carbon paste and carbon fibre electrodes.

For measurements at equilibrium, many more poorly conducting materials may be employed. This increased range of available materials is exploited in order to make the electrode materials selective to the species of interest, as in, for example, ion-selective electrodes (see Chapter 3).

External control of electrochemical cells

The current flowing through the electrical circuit can result either from the occurrence of spontaneous electrode reactions (*galvanic cell*) or from the supply of electrical energy in the external circuit which causes reactions to occur at the electrodes (*electrolytic cell*). A controlled external variable energy source (i.e. variable voltage or variable current) can permit the transition from a galvanic cell (battery) to an electrolytic cell through application of different voltages or currents.

Generally, in electrochemical research, and certainly in most electroanalytical experiments, it is the electrode process at one electrode, the *working* or *indicator electrode*, which is of prime interest. Usually, the potential of this electrode is controlled with respect to a *reference electrode* (see above) which does not pass current, by means of an instrument called a *potentiostat*; the current of the electrochemical cell passes between the working electrode and an *auxiliary electrode*. The cell thus contains three electrodes. Fig. 2.4 shows a one-compartment cell. Sometimes it is necessary to ensure that the reaction products from the auxiliary electrode cannot reach the working electrode; in these cases, a two-compartment cell is employed using, for example, a glass frit separator.

Alternatively, the current passing through the electrode/solution interface of the working electrode to the auxiliary electrode is directly controlled by means of a *galvanostat*.

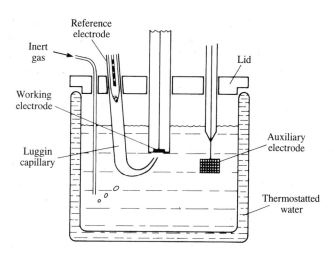

Fig. 2.4   A typical three-electrode one-compartment cell.

These instruments, which use operational amplifiers, or more recently microprocessors, enable the necessary control without the reference electrodes passing more than a few pA of current which would otherwise disturb their functioning and potential.

Liquid junction potentials

For equilibrium measurements made at zero current, an extremely important case of the Nernst equation arises from the general phenomenon of *liquid junction potentials*. These are due to the different mobilities of ions under the influence of an electric field owing to their varying sizes and charges. The different speeds of movement of cations and anions thus create a potential difference. Normally this is an unwanted contribution to the total cell potential but can be the quantity of interest, as will be seen below. The total cell potential is given by

$$E_{cell} = E_{Nernst} + E_j \tag{2.6}$$

where E_j is the liquid junction potential. Two types of liquid junction potential are commonly considered and will be described.

First, we consider a concentration cell with transport, e.g.

$$H_2, Pt \mid HCl(a_1) \mid HCl(a_2) \mid Pt, H_2 \tag{2.7}$$

where $\mid$ is the liquid junction. The partial pressures of hydrogen are equal on both sides, the only difference being in the activity of the hydrochloric acid solutions, the ions of which can migrate from one side to the other. It can be shown that

$$E_j = (t_+ - t_-)\frac{RT}{F}\ln\frac{a^\alpha}{a^\beta} \tag{2.8}$$

In this equation t_+ and t_- are the transport numbers of cation and anion, representing the fraction of charge carried by the cation and anion, respectively, from one half-cell to the other, and the superscripts α and β represent the different phases. Clearly, if the transport numbers are equal then the liquid junction potential disappears.

Electrolytes which can reduce E_j to values of 1–2 mV are potassium chloride ($t_+ = 0.49$ and $t_- = 0.51$) and potassium nitrate ($t_+ = 0.51$ and $t_- = 0.49$) and they are often used in salt bridges between two half-cells. A salt bridge is one way of providing solution contact between two independent half-cells containing solutions of different compositions; other ways are with a porous diaphragm or membrane. Sometimes, such as in the cases of the calomel and the Ag|AgCl reference electrodes, the electrolyte taking part in the reference half-cell reaction can also function simultaneously as a salt bridge.

Secondly, and for completeness, we consider the case of two solutions of different electrolytes at the same concentration with one of the ions in common. For a 1 : 1 electrolyte, the *Lewis–Sargent relationship* is obtained

$$E_j = \pm \frac{RT}{F} \ln \frac{\Lambda_\beta}{\Lambda_\alpha} \qquad (2.9)$$

where the positive sign corresponds to a common cation and the negative sign to a common anion. If the molar conductivities, Λ, of the ion that differs are equal in the two phases, then E_j will be zero.

Reconsideration of Eqn 2.8 shows that if we can create a membrane which is permeable to only one ion then the transport number of that ion will be unity and

$$E_m = \frac{RT}{z_i F} \ln \frac{a_i^\alpha}{a_i^\beta} \qquad (2.10)$$

E_m is called the *membrane potential* or *Donnan potential*. In an ideal situation, where there is no transport of any other ions, E_m changes in a Nernstian fashion with the activity of the ion in one of the phases. This is the basis of the functioning of ion-selective electrodes, and to a good approximation, of biomembranes.

2.3 Electrode processes

The increasingly widespread use of electroanalytical techniques has resulted in greater awareness of the difficulties associated with a correct interpretation of the measurements performed. On the other hand, using modern instrumentation it is often possible to obtain qualitative diagnostic and mechanistic information as well as quantitative data from the same experiment. Thus, an understanding of the basics of electrode processes, and how the various electrochemical methods can be used to extract such information, is of extreme importance.

Whereas an electrode predominantly attracts positively and negatively charged species, which may or may not undergo reaction at the surface, it should be remembered that there is also interaction with neutral species through adsorption. This makes it clear that in the description of any electrode process we have to consider the transport of species to the electrode surface as well as the electrode reaction itself. This transport can occur by *diffusion, convection* or *migration*. Diffusion is transport of species due to a concentration gradient, convection is the imposition of mechanical fluid movement and migration is the movement of species under an electric field gradient.

For electroanalytical purposes, conditions are usually created in solution so that migration of the electroactive species, i.e. the species that reacts at the electrode and which we wish to investigate, can be neglected, through the addition of a large excess of inert electrolyte (*supporting electrolyte*). Some examples of such electrolytes are shown in Table 2.2. This electrolyte transports nearly all the charge movement by ion transport from one electrode to the other through the solution and the electroactive species is left unaffected by migration effects. Thus, one is left with diffusion and convection. If there is convection, usually externally controlled, it is usually supposed to occur only as close to the surface as a distance known as the diffusion layer thickness. Closer to the surface there is only transport by diffusion. This leads to the scheme for electron transfer which is shown in Fig. 2.5.

Several important parameters and features are shown in Fig. 2.5.

1. The mass transfer coefficient, k_d, to describe diffusion from bulk solution, i.e. outside the diffusion layer, to very close to the electrode.

2. The anodic and cathodic rate constants, k_a and k_c, respectively.

3. The importance of the reagents reaching to within molecular distances from the surface of the electrode—the electrolyte double layer—and to be oriented in the correct way, in order for reaction to occur.

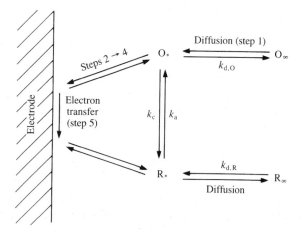

Fig. 2.5 Scheme of electron transfer at an electrode. Step 1: diffusion; Step 2: rearrangement of the ionic atmosphere; Step 3: reorientation of solvent dipoles; Step 4: alterations in central ion–ligand distances; Step 5: electron transfer.

2.4 Transport processes

Diffusion is the natural movement of species under concentration gradients from a region of high to a region of low concentration so as to annul the concentration difference (see Fig. 2.6). Diffusion is described by *Fick's first law*. In one dimension

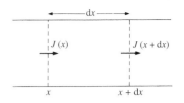

$$J = -D\frac{\partial c}{\partial x} \tag{2.11}$$

Fig. 2.6 Diffusion in one dimension, in the direction opposing the concentration gradient.

where J is the flux of species, $\partial c/\partial x$ the concentration gradient in direction x and D the proportionality coefficient known as the diffusion coefficient. Typical values for D in aqueous solution are in the range 10^{-5}–10^{-6} cm^2 s^{-1}.

The variation of concentration with time due to this movement is given by *Fick's second law*

$$\frac{\partial c}{\partial t} = D\frac{\partial^2 c}{\partial x^2} \tag{2.12}$$

In an electrode process, the concentration gradient is created by the consumption of an electroactive species at the electrode surface. A particularly interesting case is when $(\partial c/\partial t) = 0$, which means a steady-state response and, therefore, that currents do not vary with time. This situation can be achieved with hydrodynamic electrodes and with microelectrodes.

We first consider the case of a step of applied potential (potential step) at an electrode: the study of the variation of current with time for $t > 0$, called *chronoamperometry*. The potential is stepped at $t = 0$ from a value where no electroactive species react to a value where all the species that reach the electrode react (see Fig. 2.7a). For a planar electrode, which is uniformly accessible to species from bulk solution, the resulting variation of current with time calculated from Fick's second law is given by the *Cottrell equation*

$$I = nFAJ = \frac{nFAD^{1/2}c_\infty}{(\pi t)^{1/2}} \tag{2.13}$$

where I is the current measured at time t at the electrode of area A, being directly proportional to c_∞, the bulk concentration of electroactive species, as shown in Fig. 2.7b. The Cottrell equation is particularly important owing to the widespread use of pulse techniques which are based on potential steps (see Chapter 4).

At a spherical electrode of radius r_0, the relevant equation for the variation of current with time is

$$I = nFAJ = nFADc_\infty\left[\frac{1}{(\pi Dt)^{1/2}} + \frac{1}{r_0}\right] \tag{2.14}$$

which is the Cottrell equation plus a spherical term dependent on the reciprocal of the electrode radius. Thus, for small t, the first, i.e. Cottrell, term dominates and at long t the current becomes independent of time. The latter often occurs with microelectrodes ($r_0 \leq 50 \ \mu$m) under appropriate experimental conditions.

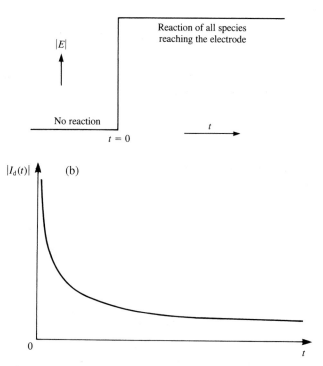

Fig. 2.7 Potential step at an electrode. (a) Scheme of application. (b) Current response to potential step as a function of time (chronoamperogram).

The concentration gradient caused by the electrode reaction tends to zero asymptotically as the distance from the electrode into solution increases. For comparison between different types of electrode and experimental methodology it is useful to define a *diffusion layer* of thickness δ, where

$$D\left(\frac{\partial c}{\partial x}\right)_0 = D\frac{(c_\infty - c_0)}{\delta} \tag{2.15}$$

where subscript '∞' refers to bulk solution and '0' to the electrode surface. When $c_0 = 0$

$$D\left(\frac{\partial c}{\partial x}\right)_0 = D\frac{c_\infty}{\delta} = k_d c_\infty \tag{2.16}$$

In this equation, k_d is *the mass transfer coefficient* $(= D/\delta)$.

Thus, the diffusion layer thickness and mass transfer coefficient corresponding to the Cottrell equation are

$$\delta = (\pi Dt)^{1/2} \quad k_d = (D/\pi t)^{1/2} \tag{2.17}$$

Another important situation arises when a current step, i.e. a step in applied current, is applied at $t = 0$ from zero and the variation of potential with time is monitored, i.e. *chronopotentiometry*. In this case the relevant expression is the *Sand equation*

$$\frac{I\tau^{1/2}}{c_\infty} = \frac{nFAD^{1/2}\pi^{1/2}}{2} \qquad (2.18)$$

The parameter $(I\tau^{1/2})$ is constant; τ is known as the transition time. The variation of potential with time, expressed by

$$E = E_{\tau/4} + \frac{RT}{nF}\ln\frac{\tau^{1/2} - t^{1/2}}{t^{1/2}} \qquad (2.19)$$

is shown in Fig. 2.8. This concept is used in potentiometric stripping analysis (Chapter 4).

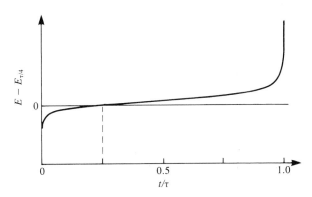

Fig. 2.8 Variation of potential with time after application of a current step at $t = 0$ (chronopotentiogram); τ is the transition time.

2.5   Kinetics of electrode reactions

Consideration of activation energies for forward and backward electrode reactions for a simple electron transfer

$$O + ne^- \rightleftharpoons R$$

led to the Butler–Volmer formulation of electrode kinetics for anodic and cathodic rate constants, respectively

$$k_a = k_0 \exp[\alpha_a nF(E - E^{\ominus\prime})/RT] \qquad (2.20a)$$

$$k_c = k_0 \exp[-\alpha_c nF(E - E^{\ominus\prime})/RT] \qquad (2.20b)$$

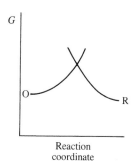

Fig. 2.9 Electron transfer energy profile for a reduction with $\alpha_c \sim 1/2$.

where k_0 is the standard rate constant, α_a and α_c are the anodic and cathodic transfer coefficients and $E^{\ominus\prime}$ is the formal potential for the system. Values of α tend to be around 0.5 for metals, reflecting the fact that the transition state is approximately halfway between reagents and products (see Fig. 2.9), although the value can vary significantly for semiconductors. However, this refers only to the rate-determining step of the electrode reaction, so that measured values of α can appear to be rather different—in the rate determining step $\alpha_a + \alpha_c = 1$.

The observed rate of reaction at the electrode surface, assuming kinetic control, is thus measured as the total current, I, where

$$I = I_a + I_c = nFA(k_a[R]_* - k_c[O]_*) \tag{2.21}$$

and the subscript $*$ refers to the reaction site very close to the electrode surface.

At equilibrium, the current is zero and so $I_a = -I_c = I_0$, where I_0 is referred to as the *exchange current*. Comparison with Eqns 2.20 shows that I_0 is directly proportional to the standard rate constant, k_0, and can therefore be used to express the rate of an electrode reaction. For example, substituting Eqns 2.20 in Eqn 2.21

$$I_0 = I_a = nFAk_0[R]_\infty \exp[\alpha_a nF(E_{eq} - E^{\ominus\prime})/RT] \tag{2.22}$$

$$= -I_c = nFAk_0[O]_\infty \exp[-\alpha_c nF(E_{eq} - E^{\ominus\prime})/RT] \tag{2.23}$$

noting that $[R]_* = [R]_\infty$ and that $[O]_* = [O]_\infty$, (the net current is zero and hence the rates of production and consumption of R are equal, the same being true for O).

Combining Eqns 2.22 and 2.23, putting $\alpha_a + \alpha_c = 1$ and rearranging

$$E_{eq} = E^{\ominus\prime} + \frac{RT}{nF} \ln \frac{[O]_\infty}{[R]_\infty} \tag{2.24}$$

which is again simply the *Nernst equation*. If $[O]_\infty = [R]_\infty = c_\infty$ then $E_{eq} = E^{\ominus\prime}$ and

$$I_o = nFAk_0 c_\infty \tag{2.25}$$

2.6   Kinetics and transport

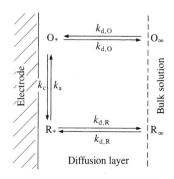

Fig. 2.10 Scheme of a simple electrode process showing diffusion (described by k_d) and kinetic (described by k_a and k_c) steps.

In real situations both kinetics and transport have to be considered together. The exponential dependence of the rate constants of an electrode reaction as described by Eqns 2.20 does not mean that currents can rise to infinite values, since limitations are imposed by the rate at which electroactive species reach the electrode surface. In Section 2.3 we assumed that transport was wholly by diffusion. The simplified scheme of Fig. 2.10 illustrates the importance of kinetics and rate of diffusion transport in how oxidation and reduction can occur.

For the purposes of this section we consider steady state, i.e. $(\partial_c/\partial t) = 0$. The observed current density, j, is then given by

$$
\begin{aligned}
j/nF &= -k_c[O]_* + k_a[R]_* \\
&= k_{d,O}([O]_* - [O]_\infty) \\
&= k_{d,R}([R]_\infty - [R]_*)
\end{aligned}
\tag{2.26}
$$

We can immediately define diffusion-limited anodic or cathodic current densities, $j_{L,a}$ or $j_{L,c}$, also referred to as limiting current densities, corresponding to consumption of all the electroactive species reaching the electrode surface. They are given by, respectively, and

$$
j_{L,a}/nF = k_{d,R}[R]_\infty \quad \text{and} \quad j_{L,c}/nF = -k_{d,O}[O]_\infty
\tag{2.27}
$$

Since, from Eqn 2.17, the ratio of the mass transfer coefficients, k_d, is related to the ratio of diffusion coefficients according to

$$
p = k_{d,O}/k_{d,R} = (D_O/D_R)^s
\tag{2.28}
$$

where $s = 1/2$ (dropping mercury electrode, stationary electrodes in stationary solution), $s = 2/3$ (hydrodynamic electrodes) or $s = 1$ (microelectrodes), it is possible to eliminate the concentrations from Eqn 2.26 and write

$$
j = \frac{k_c j_{L,c} + p k_a j_{L,a}}{k_{d,O} + k_c + p k_a}
\tag{2.29}
$$

If only O is present in solution this leads to

$$
j = \frac{k_c j_{L,c}}{k_{d,O} + k_c}
\tag{2.30}
$$

which can be rewritten as

$$
-\frac{1}{j} = \underbrace{\frac{1}{nF k_c[O]_\infty}}_{\text{kinetics}} + \underbrace{\frac{1}{nF k_{d,O}[O]_\infty}}_{\text{transport}}
\tag{2.31}
$$

Thus, when the process is transport limited, $k_c \gg k_{d,O}$ and the second term dominates (diffusion-limited current). If $k_c \ll k_{d,O}$ then

$$
-\frac{1}{j} = \frac{1}{nF k_c[O]_\infty}
\tag{2.32}
$$

and kinetics dominates the flux. The corresponding expression for an oxidation with only R in solution is

$$
-\frac{1}{j} = \underbrace{\frac{1}{nF k_a[R]_\infty}}_{\text{kinetics}} + \underbrace{\frac{1}{nF k_{d,R}[R]_\infty}}_{\text{transport}}
\tag{2.33}
$$

The values of k_a and k_c depend on the potential and on the standard rate constant, k_0, from Eqns 2.20. Two extremes can be envisaged.

(a) $k_d \ll k_0$—equilibrium is reached at the electrode surface for all values of potential. This is referred to as a *reversible reaction*. In practice, from an electroanalytical point of view, this means that no kinetic information can be extracted from the experimental data since the kinetics are too fast. If it is desired to do so, then a different technique which increases the mass transfer has to be employed so that it is no longer true that $k_d \ll k_0$.

(b) $k_d \gg k_0$—at values of potential when the species start reacting, mass transport has no influence on the rate of the process and it is completely controlled by the kinetics. However, at more extreme values of potential, and as the species close to the electrode are consumed, transport-limiting effects on the magnitude of the current begin to appear until a limiting current corresponding to total consumption of electroactive species is reached. These are referred to as *irreversible reactions*: kinetic information as well as quantitative data can be obtained.

In practice many reactions lie between the two extremes. The trick is to choose the technique such that kinetic information, if required, can be obtained easily.
We now consider reversible and irreversible systems in more detail.

Reversible reactions, $k_d \ll k_0$

For reversible reactions equilibrium will always exist at the electrode surface, and so the Nernst equation can be applied.
From Eqns 2.26 and 2.27 we can deduce

$$\frac{j}{j_{L,c}} = \frac{[O]_* - [O]_\infty}{[O]_\infty} \tag{2.34}$$

which is

$$[O]_* = \frac{j_{L,c} - j}{j_{L,c}}[O]_\infty \tag{2.35}$$

In a similar way

$$[R]_* = \frac{j_{L,a} - j}{j}[R]_\infty \tag{2.36}$$

Substituting in the Nernst equation, Eqn 2.2, (using the symbol E rather than E_{eq} to denote that here there is only equilibrium at the electrode surface and not further away from the electrode) we obtain

$$E = E^{\ominus\prime} + \frac{RT}{nF}\ln\left\{\frac{I_{L,c} - I}{I - I_{L,a}} \cdot \frac{k_{d,R}}{k_{d,O}}\right\} \tag{2.37}$$

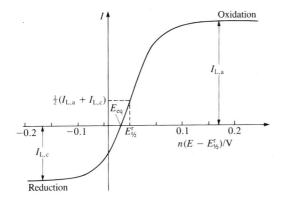

Fig. 2.11 Voltammetric profile for a reversible electrode reaction: $O + ne^- \rightleftharpoons R$ in a solution containing both O and R.

$$= E^r_{1/2} + \frac{RT}{nF} \ln \frac{I_{L,c} - I}{I - I_{L,a}} \tag{2.38}$$

where

$$E^r_{1/2} = E^{\ominus\prime} + \frac{RT}{nF} \ln \frac{k_{d,R}}{k_{d,O}} = E^{\ominus\prime} + \frac{RT}{nF} \ln \left(\frac{D_R}{D_O} \right)^s \tag{2.39}$$

using Eqn 2.28. $E^r_{1/2}$ known as the half-wave potential at which the current is exactly equal to $(I_{L,a} + I_{L,c})/2$, as shown in Fig. 2.11. A number of deductions can be made concerning reversible processes which are shown in Box 2.1.

BOX 2.1   CHARACTERISTIC PARAMETERS FOR ELECTRODE REACTIONS

Reversible reactions

- Half-wave potential, $E^r_{1/2}$, independent of $[O]_\infty$ and $[R]_\infty$

- Form of the current–potential curve independent of k_d

- Plot of $\lg[(I_{L,c} - I)/(I - I_{L,a})]$ vs. E gives a straight line of slope $0.05916/n$ V at 298 K and an intercept of $E^r_{1/2}$

Irreversible reactions

- Half-wave potential dependent on k_d

- Plot of $\lg[(I_{L,c} - I)/I]$ vs. E or $\lg[I/(I_{L,a} - I)]$ vs. E is a straight line of slope $-0.05916/\alpha_c n$ V or $0.05916/\alpha_a n$ V, respectively, with intercept $E^{irr}_{1/2}$

Irreversible reactions, $k_d \gg k_0$

The expressions for an irreversible reaction, where the kinetics are comparatively very slow, are

$$E = E_{1/2,c}^{\text{irr}} + \frac{RT}{\alpha_c nF} \ln \frac{I_{L,c} - I}{I} \tag{2.40}$$

$$E = E_{1/2,a}^{\text{irr}} + \frac{RT}{\alpha_a nF} \ln \frac{I}{I_{L,a} - I} \tag{2.41}$$

The irreversible half-wave potential is not constant but varies with the rate of transport of electroactive species to the electrode (see Fig. 2.12 and Box 2.1).

Quasi-reversible reactions

These reactions lie between the two extremes of reversible and irreversible reactions. The two waves in Fig. 2.12 partially overlap. Since, in general, only species O or R is present in bulk solution, the same equations as those for irreversible reactions can be used for quasi-reversible reactions in order to extract kinetic information.

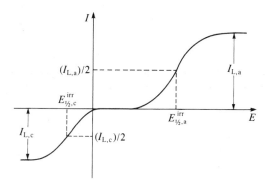

Fig. 2.12 Voltammetric profile for an irreversible electrode reaction in a solution containing both O and R. Note that the anodic and cathodic waves are completely separated.

The Tafel law

Examination of Eqns 2.20 for k_a and k_c shows that there is a region of potential close to the equilibrium potential where j depends exponentially on potential, since, here, kinetics dominates. This relationship was first discovered experimentally by Tafel. For a reduction, since $j_c/nF = -k_c[O]_*$

$$\ln |j_c| = \text{constant} - \frac{\alpha_c nFE}{RT} \tag{2.42}$$

and, in the same way

$$\ln j_a = \text{constant} + \frac{\alpha_a nFE}{RT} \tag{2.43}$$

which are expressions of the Tafel law, $\ln j \propto E$. Plots of $\ln |j|$ vs. E thus give $\alpha_c n$ and $\alpha_a n$ from the slopes (Fig. 2.13). The intercept at $E = E_{eq}$ is $\ln j_0$, where j_0 is the exchange current density, which can be related to the standard rate constant through Eqn 2.25.

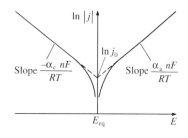

Fig. 2.13 Schematic Tafel plot showing how to measure j_0 and α.

To be more widely applicable, the Tafel law can be corrected for the effects of transport in order to enable larger sections of the voltammetric curve to be used, as long as the limiting current is measured. The corrections are included in the corresponding equations, which are, for steady-state conditions

$$\ln\left[\frac{1}{j_{L,c}} - \frac{1}{j}\right] = -\frac{\alpha_c nF(E - E^\ominus)}{RT} + \ln k_0 + \ln(nF[O]_\infty) \qquad (2.44)$$

and

$$\ln\left[\frac{1}{j} - \frac{1}{j_{L,a}}\right] = \frac{\alpha_a nF(E - E^\ominus)}{RT} + \ln k_0 + \ln(nF[R]_\infty) \qquad (2.45)$$

These equations are particularly useful as they enable values of α and k_0 to be obtained from the plots, as well as concentration determination from the limiting current value (Eqn 2.27).

2.7   The interfacial region and electrolyte double layer

The *interfacial region* is the region between the electrode and solution where electrode reactions actually occur and where the greatest potential differences across the electrical circuit appear. Considering an inert metal electrode, i.e. purely a sink or source of electrons, and according to the applied potential, a net charge will appear at the surface which will attract charge of opposite polarity. This charge separation means that there is a capacity associated with the interfacial region (often called a double layer due to the charge separation on either side of the metal/solution contact plane) which can be measured by means of impedance techniques, for example. Modern pictures of the electrolyte double layer, see Fig. 2.14, recognise that most of the surface is covered by polar solvent molecules, due to their high concentration, and also by adsorbed electrolyte ions of charge opposite to that of the surface charge— the cross-over point is the point of zero charge.

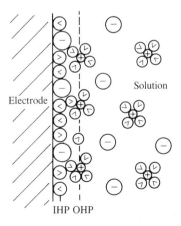

Fig. 2.14   Schematic picture of the interfacial region at an electrode, ⊖ represents a solvent molecule. IHP is inner Helmholtz plane, OHP is outer Helmholtz plane.

These electrolyte ions can be adsorbed with their solvent sheath at the outer Helmholtz plane (OHP), or in some cases after having lost their solvation at the inner Helmholtz plane (IHP), which is usually for anions (called specific adsorption)—an example is chloride ion. In order for electrode reactions to occur when the reactant is a soluble species, the reactant has to get sufficiently close to the electrode surface for electron transfer to be possible. This will involve, in general, the displacement of solvent molecules from the surface and possibly also of adsorbed ions.

It has also been recognised that the identity of the solid electrode is important in that this affects the distribution of electrons in the solid and is related to the work function of the solid, i.e. the energy needed to remove one atom from the surface of the solid to vacuum.

The importance of the interfacial region in the present context is twofold. First, it is important that the region is as thin as possible so that all the electron energy associated with the electrode can be utilised to cause the electrode reactions to occur; this is achieved by sufficiently high electrolyte concentrations. Secondly, the fact that there is charge separation in the interfacial region gives rise to a capacitative current, I_c, according to

$$I_c = C_d \frac{dE}{dt} \tag{2.46}$$

where C_d is the differential double layer capacity, which contributes to the total measured current although it is not related to the electron transfer process. For electroanalytical measurements, the capacitative contribution is usually unwanted, which means that it must be eliminated. This can be done by using, for example, a sufficiently slow rate of change of applied potential with time. Alternatively, it can be removed by subtraction, for example by subtracting a background scan conducted under identical experimental conditions but without the electroactive species in solution, relying on the fact that the electroactive species contributes only a very small fraction to the species present in the interfacial region. In some pulse techniques, a point-by-point subtraction is carried out in the way described in Chapter 4.

Typical values of the interfacial capacitance vary between 10 and 100 μF cm^{-2}.

2.8   Hydrodynamic electrodes

Imposition of controlled convection increases and defines the mass transport of solution flowing towards an electrode; this can be achieved by electrode movement or by solution movement. The diffusion layer, within which concentration gradients occur, is thinner than in the absence of convection and the response at the electrode is consequently enhanced. Electrodes placed in such conditions of solution flow are commonly known as *hydrodynamic electrodes*. If the rate of convective transport is constant, together with all other control parameters, i.e. steady state, then the electrode response is also constant. For the determination of electrode reaction and quantitative parameters, hydrodynamic electrodes are usually operated under laminar flow conditions, i.e. at convection rates where turbulence does not occur. Many of the parameters of Section 2.6 can often be easily determined with such systems.

Limiting currents

The most important parameter for characterising hydrodynamic electrodes and the region of applied potential which corresponds to greatest sensitivity is the steady-state limiting current, I_L. This will be examined for several hydrodynamic electrodes in common use. Definitions of coordinates are given in Fig. 2.15

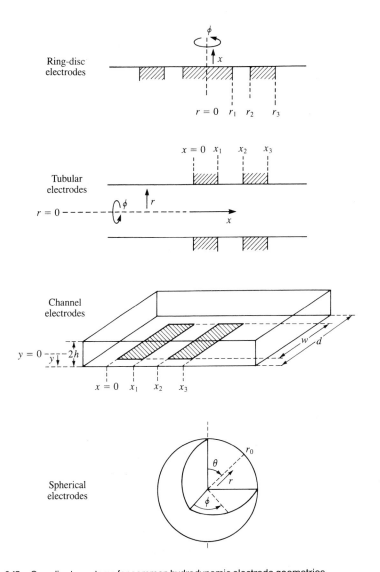

Fig. 2.15   Coordinate systems for common hydrodynamic electrode geometries.

Table 2.3   Limiting currents, I_L, at a number of common hydrodynamic electrodes[a]

Uniformly accessible electrodes[b]	
Rotating disc	$1.554nFc_\infty D^{2/3}\nu^{-1/6}\pi r_1^2 W^{1/2}$
Wall-tube	$1.22nFc_\infty D^{2/3}\nu^{-1/6}\pi r_1^2 a^{-1/2} V_f^{1/2}$
Non-uniformly accessible electrodes	
Wall-jet[b,c]	$1.59knFc_\infty D^{2/3}\nu^{-5/12}a^{-1/2}r_1^{3/4}V_f^{3/4}$
Tube	$5.43nFc_\infty D^{2/3}x^{2/3}V_f^{1/3}$
Channel	$0.925nFc_\infty D^{2/3}(h^2/d)^{-1/3}wx^{2/3}V_f^{1/3}$
Dropping mercury[d]	
Instantaneous	$709nc_\infty D^{1/2}m_1^{2/3}t^{1/6}$
Average	$607nc_\infty D^{1/2}m_1^{2/3}\tau^{1/6}$

[a] See Fig. 2.15 for explanation of coordinates.
[b] For the corresponding ring electrode expression, replace r_1^n by $(r_3^{3n/2} - r_2^{3n/2})^{2/3}$.
[c] $k = 0.86 - 0.90$.
[d] This is the Ilkovic equation which assumes uniform accessibility.
Symbols for flow parameters:

W	rotation speed (s^{-1})	V_f	volume flow rate ($cm^3\,s^{-1}$)
a	jet diameter (cm)	m_1	mass flow rate ($mg\,s^{-1}$)

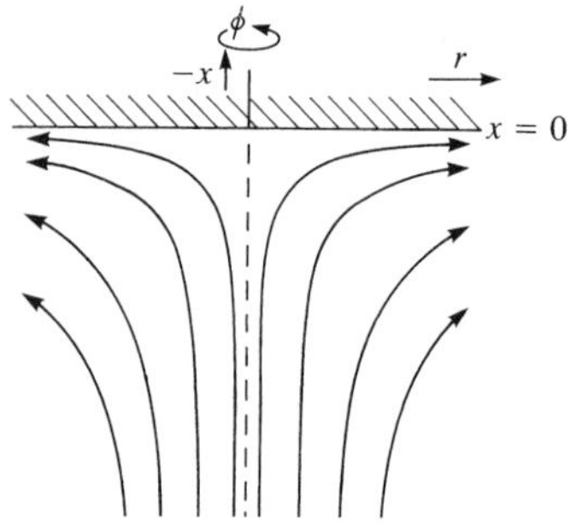

Fig. 2.16 Flow pattern at a rotating disc electrode.

Rotating disc electrode (RDE). One of the most well-known hydrodynamic electrodes is the rotating disc electrode, in which a disc of electrode material attached to a conducting rod, in order to make external electrical connection, is embedded in the centre of a coplanar mantle. The assembly is held vertically with the electrode facing downwards, attached directly to a motor or to a belt drive, and rotated about its axis in a one- or two-compartment cell of the type shown in Fig. 2.4. The flow pattern in Fig. 2.16 shows that solution is sucked up from underneath and speads out sideways. The limiting current equation, see Table 2.3, is

$$I_L = 1.554\pi r_1^2 nFc_\infty D^{2/3}\nu^{-1/6}W^{1/2} \qquad (2.47)$$

in which W is the rotation velocity in Hz, ν the kinematic viscosity and r_1 the radius of the disc electrode. As can be seen, the current is directly proportional to the electrode area, so the electrode is uniformly accessible.

The obvious limitation for widespread application of the RDE is that the electrode has to be rotated mechanically within a cell containing the solution of interest. This batch mode leads to two difficulties; first, having to change the analyte solution for each measurement series and, secondly, the possibility of depletion of electroactive species and build-up of unwanted products and intermediates. Additionally, high rotation speeds can lead to vortex formation and some turbulence; the speed at which this occurs depends on electrode and cell design.

Electrodes in flow systems. Flow systems can solve both the batch mode problems mentioned above and are more easily adapted to many real situations in the environment or in industry, since flow systems are often already operational and the electrochemical flow cell can be inserted at an appropriate point. There are a number of electrochemical flow cell detectors on the market; most are based on the wall-jet or tube/channel principles.

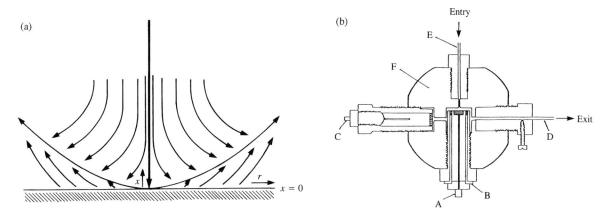

Fig. 2.17 Wall-jet electrodes. (a) Flow pattern, (b) a wall-jet electrode cell.

The wall-jet electrode, Fig. 2.17, is where a fine jet of solution impinges on the centre of a circular disc electrode and spreads out radially. The flow patterns in Fig. 2.17a show that only solution from the incoming jet can reach the electrode surface. It is important, in practice, that the radius of the jet be no more than 0.1 of the electrode radius to verify the equation in Table 2.3. Additionally, cell dimensions may not permit the establishment of the laminar hydrodynamic profile which leads to the equations shown in the table.

In the case of tube/channel electrodes there is a minimum entry length, l_e, before the steady-state Poiseuille velocity profile is verified, as shown in Fig. 2.18a. A typical channel electrode cell is shown in Fig. 2.18b.

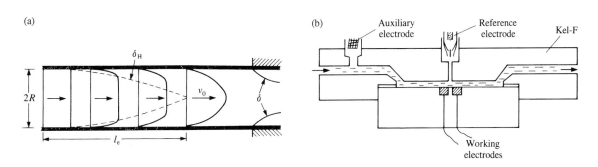

Fig. 2.18 Tube/channel electrodes. (a) Establishment of Poiseuille flow, (b) a double channel electrode cell.

Note in both these cases, as for the RDE, the dependence of I_L on $D^{2/3}$, which is common to all solid hydrodynamic electrodes (Table 2.3). However, the current is not proportional to electrode area, showing that these electrodes are not uniformly accessible, as expected from consideration of how the fluid reaches the electrode.

Flow systems can be operated in such a way that the analyte solution either flows continuously over the detector electrode or is injected into a carrier stream flowing over the electrode. In the latter case—*flow injection analysis* (FIA)—the analyte only contacts the detector for a short time and leads to a characteristic form of a peak-shaped type response, similar to that of an HPLC detector. FIA reduces the amount of analyte employed and can permit the addition of appropriate reagents at the correct points in the carrier stream, as well as switching from one stream to another (see scheme in Fig. 2.19); suitable automation can permit the analysis of many samples per hour. However, the continuous flow technique is simpler and may be better adapted for analyses of, for example, process streams, effluents or river water, where the continuous flow already exists and can be exploited for *in situ* analysis.

A hybrid approach is *batch injection analysis* (BIA), in which discrete samples are injected from a micropipette directly over the centre of an electrode immersed in stationary electrolyte solution. During the injection, the electrode response has all the characteristics of a continuous flow system.

Sonotrodes. The application of ultrasound in the zone of the electrode surface increases the mass transport through turbulence and cavitation but is equivalent to convection and leads to a steady-state response. The equivalence to hydrodynamic electrodes is also demonstrated by the same $D^{2/3}$ dependence in the experimentally measured current as for the other solid electrodes shown in Table 2.3. Although there is an associated inherent noise level, the advantage of ultrasound is that it can reduce blocking of the electrode surface by adsorption of reaction products and lead to a lesser decrease in electrode response with time.

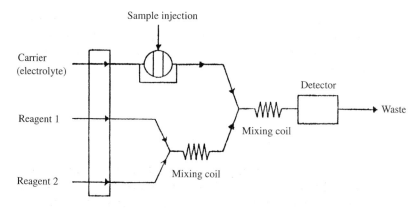

Fig. 2.19   Example of a three-channel FIA manifold with sample injection.

Dropping mercury electrode. The dropping mercury electrode (DME) can be regarded as a hydrodynamic electrode due to its cyclic operation. A continuous flow of mercury from a reservoir through a capillary leads to formation of mercury drops which grow in size until the drop falls, owing to the effect of gravity (the drop time, τ). The scheme of operation and variation of current with time are shown in Fig. 2.20. As seen from the limiting current equation in Table 2.3, the current depends on $m_1^{2/3}$, where m_1 is the mass flow rate, the value of which can be altered by changing the height of the reservoir, and on $D^{1/2}$ (for solid hydrodynamic electrodes it is $D^{2/3}$). It is normal practice nowadays to knock the capillary mechanically to remove the mature drop so as to ensure better reproducibility of the drop time.

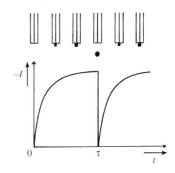

Fig. 2.20 Scheme of operation of a dropping mercury electrode and variation of current with time.

Besides dropping mercury electrodes, hanging mercury drop electrodes (HMDE) and static mercury drop electrodes (SMDE) are used; they both have a constant drop size. In the former case, one sessile drop is used throughout the experiment. In the latter, the drops are periodically renewed but the apparatus is designed so as to make a fixed-size (and reproducible) drop very quickly by a short application of high pressure on the mercury in the capillary, and to hold it for the specified period, after which it is mechanically dislodged and replaced with a new drop.

Double hydrodynamic electrodes

Most solid hydrodynamic electrodes can exist in a double electrode version, with two independent working electrodes along the direction of solution flow, a *generator electrode* (upstream) and a *detector electrode* (downstream), the potential of which is controlled independently using a *bipotentiostat*. For disc electrodes, the equivalent is a downstream ring electrode concentric with the disc and separated from it by a small insulating gap. The cell also contains an auxiliary electrode and a reference electrode, making a total of four electrodes.

Double electrodes can be of importance:

- To measure, at the detector electrode, a product arising from the electrode reaction at the generator electrode, possibly after reaction in homogeneous solution or following partial decomposition. This is described by the collection efficiency, N.

- To remove an 'unwanted' electroactive species at the generator electrode that would otherwise interfere with the required determination, which can then be carried out at the downstream electrode. This is described by the shielding factor.

The simplest reaction scheme is

$$\begin{aligned} \text{generator} \quad & A \pm n_1 e^- \rightarrow B \\ \text{detector} \quad & B \pm n_2 e^- \rightarrow C \end{aligned}$$

where C can be equal to A, implying regeneration of the reactant. Some B will escape from the diffusion layer into bulk solution before reaching the detector electrode. The *collection efficiency*, N, the fraction of B which reaches the detector is defined by

$$N = \left| \frac{n_1 I_{\text{det}}}{n_2 I_{\text{gen}}} \right| \qquad (2.48)$$

If there are no reactions in solution this is the *steady-state collection efficiency*, N_0, and has the form described in Box 2.2.

Note that N_0 is described in terms of α and β, which are themselves purely a function of electrode geometry; this independence on convection rate is an extremely useful simplification. Kinetic collection efficiencies, when the product of reaction at the upstream electrode undergoes a homogeneous reaction in solution, can also be derived.

BOX 2.2   COLLECTION EFFICIENCIES AT DOUBLE HYDRODYNAMIC ELECTRODES

Steady-state collection efficiency, N_0

$$N_0 = 1 - F(\alpha/\beta) + \beta^{2/3}[1 - F(\alpha)]$$
$$- (1 + \alpha + \beta)^{2/3}\{1 - F[(\alpha/\beta)(1 + \alpha + \beta)]\}$$

where

$$F(\theta) = \frac{3^{1/2}}{4\pi} \ln\left[\frac{(1 + \theta^{1/3})^3}{1 + \theta}\right] + \frac{3}{2\pi} \arctan\left(\frac{2\theta^{1/3} - 1}{3^{1/2}}\right) + \frac{1}{4}$$

and α and β are given by*

	α	β
Rotating ring-disc electrode (RRDE)	$\left(\frac{r_2}{r_1}\right)^3 - 1$	$\left(\frac{r_3}{r_1}\right)^3 - \left(\frac{r_2}{r_1}\right)^3$
Wall-tube ring-disc electrode (WTRDE) / Wall-jet ring-disc electrode (WJRDE)	$\left(\frac{r_2}{r_1}\right)^{9/8} - 1$	$\left(\frac{r_3}{r_1}\right)^{9/8} - \left(\frac{r_2}{r_1}\right)^{9/8}$
Tube double electrode (TDE) / Channel double electrode (CDE)	$\left(\frac{l_2}{l_1}\right) - 1$	$\left(\frac{l_3}{l_1}\right) - \left(\frac{l_2}{l_1}\right)$

* See Fig. 2.15 for explanation of electrode dimensions

The *shielding factor*, describing the removal of interfering electroactive species, can vary between 0 and 1 and is determined by the ratio of the limiting current at the detector electrode with the generator electrode connected, $I_{\text{L,det}}$, to the limiting current with it disconnected, $I^0_{\text{L,det}}$

$$I_{\text{L,det}}/I^0_{\text{L,det}} = (1 - N_0\beta^{-2/3}) \qquad (2.49)$$

where β is a geometric parameter, see Box 2.2. Thus, in order to remove the unwanted electroactive species to the greatest extent, it should be arranged that

$N_0\beta^{2/3} \to 1$, corresponding to a small interelectrode gap and a thin detector electrode.

Note that there are also cells with two (or more) working electrodes in which both electrodes are side-by-side in parallel relative to the solution flow. Assuming they are not too close to each other they are completely independent. Applying different applied potentials can lead to multicomponent detection.

2.9 Microelectrodes

Microelectrodes are electrodes which have at least one dimension in the micrometre range and thus have particular characteristics which are a direct function of their small size. Microelectrodes can have many different forms. Some are illustrated as cross-sections in Fig. 2.21, which also shows, schematically, the way in which diffusion of electroactive species towards the electrodes occurs.

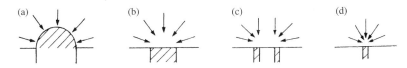

Fig. 2.21 Types of microelectrode: (a) hemisphere, (b) disc, (c) ring, (d) band.

Reduction of size leads to lower capacitative contributions to the total current and the possibility of attainment of steady-state currents within a short time, as indicated by Eqn 2.14. At a hemispherical electrode, the steady-state current in the limiting current region is

$$I = \frac{nFADc_\infty}{r_0} = 2\pi nFr_0Dc_\infty \qquad (2.50)$$

This equation can be rewritten in terms of the surface length, d, where $d = \pi r_0$ (the distance over the electrode surface from one point at the edge of the hemispherical electrode to the equivalent point on the opposite side) as

$$I = 2nFdDc_\infty \qquad (2.51)$$

At a plane disc microelectrode, one obtains the same equation, Eqn 2.51, with only a small error, where the surface length is now given by $d = 2r_0$. Note that, unlike solid electrodes under forced convection conditions, the exponent of D is 1 rather than 2/3. This should be borne in mind when employing equations such as Eqn 2.41 for the determination of formal potentials of electrode reactions. They are very insensitive to forced convection in solution.

There are several general advantages of microelectrodes besides obtaining steady-state currents after short times, usually within 0.1 s, without convection. These are, first, the small size, which enables them, in the form of fibres for example, to be inserted in places where other electrodes are too large, secondly, the high current density which leads to good signal resolution and low detection limits, and, thirdly, the low total current. This last means that they can be employed in more highly resistive media than macroelectrodes—often without the addition of any inert electrolyte.

2.10   Electroanalytical titrations

The essence of any titration is to determine the concentration of the substance (solution) titrated by adding quantities of a solution of known concentration until there is stoichiometric equivalence, denoted by some change in a property. Electroanalytical titrations make use of the concepts which have been outlined in this chapter in a simple way; they can be performed potentiometrically, amperometrically, coulometrically or conductimetrically.

It is not the objective of this book to describe the practical details of electroanalytical titrations. Coulometric titrations are based on the coulometrically monitored generation of a species which reacts with an analyte in solution until it is completely consumed. Conductimetric titrations rely on measurements of the change of total solution conductivity but are not species selective. However, potentiometric and amperometric titrations can exhibit good selectivity and will be described, given their use in routine electroanalysis.

Potentiometric titrations

Potentiometric titrations are based on the measurement of the equilibrium potential of a solution at an indicator electrode with respect to a reference electrode. Measurements are made at zero current, which with modern instrumentation means using a high impedance digital voltmeter. For the titration to give accurate results, the potential must change abruptly in the region of the equivalence point. In order to ensure this in solutions containing various ions, ion-selective electrodes can be usefully employed such as, for example, pH electrodes for acid–base titrations. The equivalence point can be determined from the sharp variation of E in the plot of E vs. volume of titrant, from the first or second derivative of the titration curve or from a Gran plot (see Fig. 2.22).

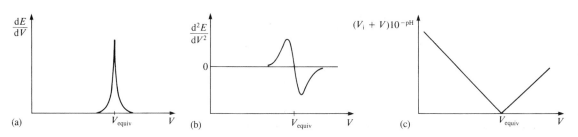

Fig. 2.22   Methods for equivalence-point detection from potentiometric titration curves: (a) first derivative; (b) second derivative; (c) Gran plot for titration of strong acid (volume V_i) with strong base.

The Gran method involves the transformation of the titration curve, the shape of which is governed by the Nernst equation, into two straight lines which intersect at the equivalence point, at potential E_{equiv}. The great advantage is that a large section of the whole titration curve is used, which can significantly increase precision.

After addition of half the volume of titrant necessary to reach the equivalence point the potential is equal to the formal potential, $E_1^{\ominus\prime}$ of the titrated species. The formal potential of the other couple, $E_2^{\ominus\prime}$, i.e. the titrant, can also be determined through

$$E_{equiv} = \frac{n_1 E_1^{\ominus\prime} + n_2 E_2^{\ominus\prime}}{n_1 + n_2} \qquad (2.52)$$

A variation on these titrations is the *bipotentiometric titration*, in which a current, usually of 5–10 μA, is passed between two Pt electrodes and the potential difference between them is monitored. At the equivalence point a sharp change in potential difference occurs.

Amperometric titrations

Amperometric titrations are based on the measurement of the current when a fixed potential difference is applied between an indicator (redox) electrode and a reference electrode; unlike potentiometric titrations, in this case the reference electrode also passes current. The potential difference is applied nowadays by means of a potentiostat, and the variation of current on adding titrant is monitored. Since the current that flows is directly proportional to concentration, the indicator electrode potential is chosen so as to monitor the concentration of titrated species (which decreases to zero at the equivalence point) or of the titrant (which increases from zero at the equivalence point) or both.

Biamperometric titrations differ from amperometric titrations in that both electrodes are redox electrodes. This means that the potential of both can vary during the course of the titration, whereas in ordinary amperometric titrations the potential of one of the electrodes, the reference electrode, is fixed. No current flows at the beginning of the titration or at the equivalence point, since at these points neither the titrated nor titrant redox couple has both oxidised and reduced species present in analyte solution (Fig. 2.23).

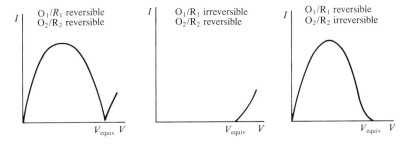

Fig. 2.23 Equivalence point determination in biamperometric titrations.

Amperometric titrations can also be carried out with a three-electrode system controlled by a potentiostat. The current at the working electrode to which a constant value of potential is applied corresponding to electrode reaction of the titrated species or titrant is monitored. Many variations on this theme can be imagined.

3 Potentiometric sensors

3.1 Principles of potentiometric sensors

The goal of this chapter is to describe the functioning of potentiometric sensors, how they can be usefully employed and in which experimental situations; although some mention of potentiometric measurements has already been made in the previous chapter.

Potentiometric sensors work through the measurement of an equilibrium potential, i.e. the potential at zero current, of the sensor vs. a suitable reference electrode. These potentials are a function of the activity of the species in solution, not of their concentration. The Debye–Hückel equation relates concentrations to activities and can often be employed; indeed, potentiometric measurements can be used to test the Debye–Hückel theory. Only for dilute solutions is it reasonable to assume that activity and concentration are equal.

Clearly, for these sensors to be useful they must have a sufficiently fast response and be sufficiently selective in media containing various species, besides having a sufficiently good detection limit.

3.2 Experimental set-up and instrumentation

In designing a suitable experimental set-up (Fig. 3.1) two situations can be imagined.

1. The potential is determined by oxidation–reduction equilibria (through the Nernst equation) at the electrode–solution interface of a redox indicator electrode, such as platinum. The difficulty which can arise is if both oxidised and reduced species of more than one redox couple are present in solution, so that they contribute to the overall equilibrium potential, which is thus a mixed potential. Thus, such a measurement can have low selectivity in real situations.

2. The potential measured is the difference of potential across a membrane, which is influenced by the activity of the species on either side of the membrane. In *selective electrodes*, most of them *ion-selective electrodes* (ISEs), the membrane is selective to the species of interest and is thus the selective component of the electrode.

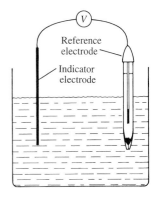

Fig. 3.1 Experimental arrangement for potentiometric measurements.

Currently, potentiometers are rarely employed and high input impedance voltmeters are used instead. These are not the simple digital voltmeters widely available which have input impedances of the order of 1 MΩ (which means that for a potential difference of 1 V a current of 1 μA has to flow) but have at least a 10^{12} Ω input impedance.

Although the basic set-up is extremely simple, problems can arise from electrical noise. Thus, electrical shielding is extremely important. Signal amplification is another route to increasing the signal/noise ratio, which can be done either using an amplifier close to the electrode itself or directly through integrated circuit technology.

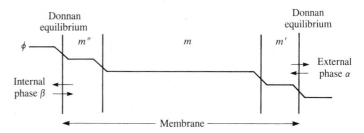

Fig. 3.2 Scheme showing operation of a glass electrode.

3.3 Functioning of ion-selective electrodes

Ion-selective electrodes (ISEs) are based on the potential difference created by the movement of ions between a solution phase, α, and a membrane phase, β, (solid or liquid in a support matrix) (see Section 2.1) and exemplified in Fig. 3.2 for a glass electrode. The membrane potential, E_m, is then given by

$$E_m = \frac{RT}{z_i F} \ln \frac{a_i^\alpha}{a_i^\beta} \qquad (3.1)$$

for an ion i of charge z_i. If the activity of the ion i in the membrane remains constant then

$$E_m = \text{constant} + \frac{RT}{z_i F} \ln a_i^\alpha \qquad (3.2)$$

and a variation of $59/z_i$ mV in E_m per decade of variation in activity should be observed at 298 K. In this way activities in the range $10^{-6} - 10^{-1}$ can be measured.

The membrane potential is measured by means of the potential difference between an internal and an external reference electrode. Thus the membrane is the link between two halves of a concentration cell (which may not be symmetrical).

A perfect ISE responds to only one ion in solutions containing 'any' ion. In practice, this ideal situation cannot be met, particularly when ions with similar characteristics are present, such as occurs with mixtures of halide ions. The interference effects of other ions depend on the potentiometric selectivity coefficient, $K_{i,j}^{pot}$. If a linear concentration gradient within the membrane is assumed then the relevant equation, the Nicolsky–Eisenman equation, for the potential difference is

$$E = \text{constant} + \frac{RT}{z_i F} \ln\left(a_i + \sum_j K_{i,j}^{\text{pot}} a^{z_i/z_j}\right) \quad (3.3)$$

where j are the interfering species. Since, in Eqn 3.3, it is the product $K_{i,j}^{\text{pot}} a^{z_i/z_j}$ which determines the extent of interference, low activities of interfering ions can mean no interference effects even if $K_{i,j}^{\text{pot}}$ is reasonably large. These selectivity coefficients can be measured by two procedures.

(a) Fixed interference method, in which the potential of a cell is measured with solutions containing a constant level of interferent and varying primary ion activity—extrapolation to zero activity of the primary ion leads to the selectivity coefficient of the interferent.

(b) Separate solution method, in which the potential in two solutions is measured, one containing the primary ion only and the other containing the interferent ion only, but at the same activity.

Of these two procedures the first is better since in real situations the two ions will coexist in the same solution; indeed, different values of selectivity coefficient can be obtained by the two methods. Other problems occur if interfering ions exhibit a non-Nernstian response and if ions of unequal charge are involved, so that Eqn 3.3 cannot be applied. In fact, the concentration gradient throughout the whole membrane is unlikely to be linear, but Eqn 3.3 provides a convenient way of measuring the influence of interferents.

3.4 Types of selective electrode

Selective electrodes are divided into three classes:

(a) primary ion-selective electrodes, encompassing crystalline and non-crystalline electrodes;

(b) compound or multiple membrane ion-selective electrodes, including gas-sensing electrodes and enzyme substrate electrodes;

(c) all-solid state ion-selective electrodes, depending on both electronic and ionic conductivity in order to function.

In the following sections examples of these types of electrode will be introduced.

Glass electrodes

Glass electrodes are an example of non-crystalline electrodes; they were the first ISEs to be developed and are used mainly to measure pH. Glass is an amorphous solid consisting principally of silicates and is permeable to H^+ and to Na^+ and K^+. The composition of the glass determines the permeability to each type of ion, but some interferences will always occur.

The functioning of this electrode is by exchange of protons from solution with sodium ions at the surface region to a depth of about 50 nm

$$H^+_{soln} + Na^+_{surf} \leftrightarrow H^+_{surf} + Na^+_{soln}$$

Transport across the membrane is wholly by alkali metal cations, usually Na^+. It is thus clear that for low proton and high alkali metal ion concentrations in solution this exchange is not complete and the potential is higher (the pH is lower) than expected. Typical values for the sodium potentiometric selectivity coefficient are around 10^{-12}. The relevant form of Eqn 3.3 is

$$E = \text{constant} + 0.0592 \lg(a_{H^+} + K^{pot} a_{Na^+}) \tag{3.4}$$

which is known as the *Eisenman equation*.

In strong acid solutions, apart from the fact that acid attacks the glass surface, the activity coefficient of the hydrogen ion can depend very much on the environment and this can lead to deviations in the potential. Deviations at high activities are a general problem common to all ion-selective electrodes.

In a pH measurement, the potential difference between two reference electrodes on either side of the glass membrane, in the form of a bulb, is monitored. In general, the internal and external reference electrodes are the same, usually Ag|AgCl, although external SCEs are also employed. A typical cell could thus be

$$\underbrace{Ag \mid AgCl \mid HCl(0.1M)}_{\text{internal reference}} \mid \text{membrane} \mid \text{test solution} \;\vdots\; \underbrace{KCl(3M) \mid AgCl \mid Ag}_{\text{external reference}}$$

Often, nowadays, the two reference electrodes are combined together with the glass membrane into one cylindrical package in order to facilitate practical manipulation, as shown in Fig. 3.3.

It is of crucial importance always to calibrate the glass electrode before measurements are made. This is partly because an asymmetry in the potential arises between both sides of the membrane due to differences in the structure and composition of the two glass surfaces. The asymmetry potential can also change with time when the electrode is used, so that periodic calibration is extremely necessary. Eventually, and especially if the electrode is not maintained in good condition by careful cleaning and storage when not in use, such phenomena can make the electrode unusable, although the lifetime can be increased by following the manufacturers' regeneration procedures.

Either one-point or two-point calibration of pH glass electrodes, also applicable to other ISEs, can be carried out (Fig. 3.4). One-point calibration means assuming a value for the variation of potential with activity. For pH this is

$$E = \text{constant } 0.0592 \text{ pH} \tag{3.5}$$

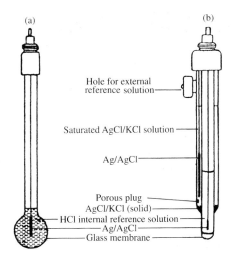

Fig. 3.3   Typical design of (a) a simple and (b) a combined glass pH electrode.

Hole for external
reference solution

Saturated AgCl/KCl solution

Ag/AgCl

Porous plug
AgCl/KCl (solid)
HCl internal reference solution
Ag/AgCl
Glass membrane

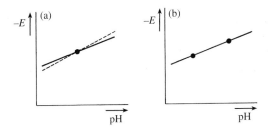

Fig. 3.4   Methods for calibrating ion-selective electrodes, illustrated for a pH electrode. The solid line refers to the variation in practice, the dotted line to the theoretical variation (0.0592 V per unit of pH at 25°C). (a) One-point calibration, (b) two-point calibration.

and calibration gives the value of the constant, so that for an unknown solution of pH_1

$$pH_1 = pH_{buff} + \frac{E_{buff} - E_1}{0.0592} \qquad (3.6)$$

Although this sometimes suffices, the accuracy is worse the greater the difference between the calibrated and unknown pH values. Two-point calibration also provides the real slope, S, according to

$$pH_1 = pH_{buff} + \frac{E_{buff} - E_1}{S} \qquad (3.7)$$

Naturally, calibration buffers should be chosen which encompass the range of pH of the unknown solution if possible. Thus, for measurement in acid solution, buffers of pH 7 and 4 are often used, and for measurements in

Table 3.1   National Institute of Standards and Technology buffer solutions

Composition	pH at 25°C
KH tartrate, saturated at 25°C	3.557
KH_2 citrate, 0.05 *m*	3.766
KH phthalate, 0.05 *m*	4.008
MOPSO [a], NaMOPSO, NaCl, all 0.08 *m*	6.865
KH_2PO_4, Na_2HPO_4, both 0.025 *m*	6.865
HEPES [b], NaHEPES, NaCl, all 0.08 *m*	7.516
KH_2PO_4 0.008695 *m*, Na_2HPO_4, 0.03043 *m*	7.413
$Na_2B_4O_7$ 0.01 *m*	9.180
$NaHCO_3$, Na_2CO_3, both 0.025 *m*	10.012

[a] (3-*N*-morpholino)-2-hydroxypropanesulfonic acid
[b] (*N*-2-hydroxyethylpiperazine-*N'*-2-ethanesulfonic acid

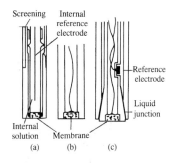

Fig. 3.5   Solid-state membrane ion-selective electrodes with, (a) internal reference electrode, (b) internal ohmic contact, (c) internal ohmic contact and combined external reference electrode.

alkaline solution, buffers of pH 7 and 9. Standardised buffer solutions are given in Table 3.1, chosen because of their very well-defined and stable pH value; note that the pH changes with temperature.

Crystalline membrane electrodes

As the name implies, the sensorial part of this type of selective electrode is a solid-state crystalline membrane (see Fig. 3.5). A *homogeneous membrane* is an ionic solid with low solubility product, the sensored ion being either the cation or the anion of the solid. The potential arises because of exchange between the solution and surface of the ionic crystal; transport through the crystal is via migration of defects in the crystal structure.

Some examples of homogeneous membrane electrodes are shown in Table 3.2. Perusal of the table shows that, in principle, the electrode will respond to both cation and anion of the solid membrane, so that the main 'interfering'

Table 3.2   Crystalline homogeneous membrane electrodes

Primary ion	Active material	Major interferences
F^-	LaF_3	OH^-
Cl^-, S^{2-}	AgCl	Br^-, I^-, CN^-, NH_3, $S_2O_3^{2-}$, S^{2-}
Br^-	AgBr	I^-, CN^-, NH_3, $S_2O_3^{2-}$, S^{2-}
I^-	AgI	CN^-, S^{2-}
CN^-	AgI	I^-, S^{2-}
SCN^-	$AgSCN + Ag_2S$	S^{2-}, I^-, Br^-
S^{2-}	Ag_2S	Ag^+
Ag^+	Ag_2S	S^{2-}
Cd^{2+}	CdS (or $CdS + Ag_2S$)	H^+, Mn^{2+}, Pb^{2+}, Fe^{3+}, $Cr_2O_7^{2-}$
Pb^{2+}	$PbS + Ag_2S$	Cu^{2+}, Cd^{2+}

species in the determination of the cation or anion is the corresponding anion or cation from the membrane, respectively. The electrode is also sensitive to other ions that can form sparingly soluble precipitates on the membrane surface, particularly if they have lower solubility products than the material of the membrane. As an example, the silver chloride electrode responds to Br^- and I^-.

A practical difference from glass electrodes is the fact that the internal reference electrode (Fig. 3.5a) is often replaced by an ohmic contact (Fig. 3.5b). A schematic cell for the measurement of Ag^+ could thus be

$$Ag \mid AgCl\mid KCl(0.1M) \vdots \text{test solution containing} \mid AgX \text{ membrane} \mid \underset{\text{ohmic contact}}{Ag}$$
$$Ag^+ \text{ but not } X^-$$

Another practical strategy is to incorporate small crystals of the ionic crystal in a plastic, inert matrix to form a *heterogeneous membrane* electrode; such matrices might be poly(vinylchloride) (PVC) or silicone rubber. Alternatively, the active substance can be placed on hydrophobised graphite or conducting epoxy resin. In either case the resulting package is more flexible and thus more resistant to breakage.

Non-crystalline membrane electrodes

Non-crystalline membrane electrodes consist of a polymeric support matrix of a material such as PVC or silicone rubber which contains a solvent and ion exchanger, usually a chelating agent, selective to the species to be determined. The basic design is shown in Fig. 3.6. They are essentially of two types—ion exchange and neutral carrier. Transport across the membrane is by exchange of the species of interest between adjacent chelating agents, which are usually macrocycles. Glass electrodes (see above) are an example of a non-crystalline membrane electrode. Some other types will now be described.

Ion exchange membranes. These are known as charged, mobile-carrier electrodes and can be subdivided depending on whether the electrode is sensitive to anions or cations. The chelating agent is hydrophobic and has the opposite polarity. Examples are given in Table 3.3.

Neutral carrier membranes. In these electrodes the chelating agent has no charge and complexes the ion of interest selectively, or membranes containing a hydrophobic ion pair dissolved in plasticised polymer. One of the most well known is the potassium electrode using the macrocycle valinomycin as neutral carrier; crown ethers in general are good candidates for the active membrane component. Other examples are given in Table 3.4.

Gas-sensing electrodes

Gas-sensing electrodes are examples of multiple membrane ISEs: they are simply selective electrodes, the membrane of which is covered by a second gas-permeable membrane which only allows certain molecules to pass. In

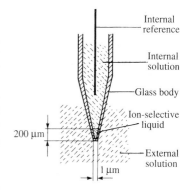

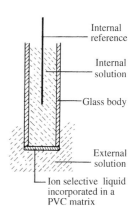

Fig. 3.6 Non-crystalline membrane ion-selective electrodes.

Table 3.3   Examples of charged mobile-carrier electrodes

Primary ion	Active material	Major interferences
Positively charged		
Cl^-, Br^-, I^-	Cetyl-$(CH_3)_3$NOH in octanol	$ClO_4^-, NO_3^-, OH^-, SO_4^{2-}$
I^-	Aliquat-336S on coated wire	NO_3^-
SCN^-	Aliquat-336S on coated wire	I^-, SO_4^{2-}
NO_3^-	Cetyl-$(CH_3)_3$NOH in	
	octyl-2-nitrophenyl ether	$Cl^-, Br^-, SO_4^{2-}, NO_2^-$
Negatively charged		
Ca^{2+}	Didecylphosphate in	
	di-*n*-octylphenylphosphonate	H^+, Ba^{2+}, Mg^{2+}
Zn^{2+}	Zn salt of di-*n*-octyl-phenyl	$Mg^{2+}, Ca^{2+}, Sr^{2+}, Cd^{2+},$
	phosphoric acid	Pb^{2+}, Ba^{2+}

Table 3.4   Neutral-carrier membrane electrodes

Primary ion	Active material	Major interferences
K^+	30-crown-10 derivative in PVC	Rb^+, Cs^+, Ca^{2+}
K^+	Valinomycin in diphenyl ether	NH_4^+, Rb^+, Cs^+
Ca^{2+}	Antibiotic A-23187	Sr^{2+}, Na^+, Mg^{2+}
Ba^{2+}	Polyethyleneglycol derivative	Sr^{2+}

most important examples a small amount of electrolyte is placed between the selective membrane (typically a pH glass membrane) and the outer membrane whose pH is altered according to the partial pressure of the gas. Some examples are given in Table 3.5.

Table 3.5   Examples of gas-sensing electrodes

Species	Lower limit (M)	Sensor ion	Inner solution
CO_2	10^{-5}	H^+	0.01 M $NaHCO_3$
NH_3	10^{-6}	H^+	0.1 M NH_4Cl
H_2S	10^{-8}	S^{2-}	Buffer, pH 5
Cl_2	10^{-3}	Cl^-	Buffer, HSO_4^-

Potentiometric enzyme electrodes

Enzymes are highly specific catalysts. A product of an enzyme reaction can often be indirectly monitored at a selective electrode, in the same way as for gas-sensing electrodes. For this purpose, the enzyme has to be immobilised on

Table 3.6 Potentiometric enzyme electrodes

Species	Enzyme	Sensor
L-Arginine	Arginine carboxylase	CO_3^{2-}
Glucose	Glucose oxidase	
	(produces H_2O_2)	I^-
Amygdaline	β-glucosidase	CN^-
Urea	Urease	NH_3
NO_2^-	Nitrite reductase	NH_3

a membrane which is placed over the electrode, or possibly coated directly on to it. A pertinent example of these *potentiometric biosensors* is the degradation of urea by urease leading to the formation of ammonium ion which is detected at the ammonium-selective neutral-carrier electrode (see Table 3.5). Other examples are found in Table 3.6.

3.5   Miniaturisation strategies

The electrodes described above can only be miniaturised up to a certain point, due to difficulties in fabrication. The first step in this direction is to replace the internal reference solution by an electronic conductor (ohmic contact). Other approaches have been developed with varying success.

Coated-wire electrodes

One strategy tried with some success, but which showed difficulties with respect to reproducibility, was that of coated-wire electrodes. Here, the membrane material is applied directly on to a metal wire. This assembly has no internal reference electrode, which is replaced by an ohmic contact. Despite their disadvantages, these electrodes are cheap and disposable.

Ion-selective field-effect transistors

An alternative approach, following the same line of reasoning as coating a conducting wire, but achieving reproducible signals with a high signal/noise ratio, is through semiconductor transistor technology. The function of a conventional field effect transistor is to respond to tiny voltage differences in a metallic gate between source and drain and convert them into low impedance output signals in the form of currents.

In the ion-selective field-effect transistor (ISFET), the metallic gate is replaced with an ion-selective membrane which is in contact with the solution. In this way the output (drain) signal, I_d, can be directly related to the activity of ions in solution such as protons. The scheme of operation is shown in

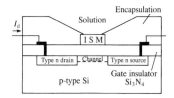

Fig. 3.7 Scheme of operation of an ISFET.

Fig. 3.7. Other important applications relying on similar technology are chemically sensitive FETs (CHEMFETs), enzyme FETs (ENFETs) and immunosensitive FETs (ImmunoFETs).

3.6 Criteria for choosing a potentiometric sensor

When employing potentiometric sensors, a number of criteria must be considered. These include the answers to the following questions.

(a) Is the selectivity sufficient for the application, i.e. what are the interfering species and what are their activities/concentrations?

(b) Does calibration lead to a straight line variation with activity of the species of interest? Is this variation (approximately) Nernstian? Does this vary with electrode ageing?

(c) Is the detection limit sufficiently low? Fig. 3.8 shows how detection limits are determined following the recommendation of IUPAC, corresponding to the intercept between the linear response and zero response for very low activities.

(d) Is the response time of the electrode to a jump in activity sufficiently small (usually defined as the time needed to reach 95% of the final value)?

(e) Is the lifetime of the electrode sufficiently long in continuous use or when stored without use?

(f) Can the electrode surface be renewed/regenerated if necessary?

These sorts of questions lead to different answers depending on the species to be analysed, the medium where the measurements are to be carried out and the precision and accuracy required.

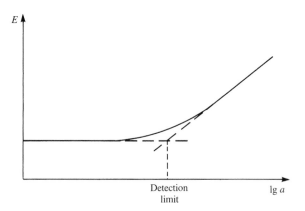

Fig. 3.8 Determination of detection limits at a potentiometric sensor according to IUPAC recommendations.

Particular problems can arise in the analysis of flowing solutions, such as are encountered in flow injection systems or in HPLC detection systems, owing to the solution movement. These factors include:

1. Positioning of the external reference electrode. Local electric fields between indicator and reference electrodes can influence signal stability and can be increased by solution flow. Therefore, the electrodes must be placed as close as possible.

2. An increase in electrode response time.

3. Deterioration of the membrane due to removal of species from the surface, leading to shorter electrode lifetime and the necessity of more frequent calibration. Additionally, an inferior response can occur due to increased adsorption of suspended matter on the electrode surface. Solid residues should be removed by filtering and solution conditions adjusted to optimise the response.

4 Voltammetric sensors

4.1 Principles of voltammetric sensors

Voltammetric sensors function by measurement of the current response as a function of applied potential (or the potential response as a function of applied current); in other words, they depend on the registering of current–potential profiles. A special case of voltammetric sensors is *amperometric sensors* where a fixed potential is applied and the current is registered. The recording of current as a function of time (*chronoamperometry*) can give important and useful information, as can the charge passed—*coulometry* and *chronocoulometry*. For this purpose, it is necessary that the species of interest is electroactive at the electrode material at a reasonable value of potential where neither solvent nor electrolyte decomposition occur (see Section 2.2).

Hydrodynamic electrodes are important voltammetric sensors and were introduced in Chapter 2. They are usually operated under steady-state conditions so that steady-state voltammetric curves can be recorded. For purely analytical purposes, it is clear that the highest sensitivity amperometric sensors are those where a potential is applied corresponding to the limiting current region (see Figs 2.11 and 2.12).

This chapter will concentrate mainly on describing the basic principles of a number of different types of non-steady-state techniques and their use in analysing electroactive species in solution, together with more elaborate schemes for application to detection of trace species.

4.2 Experimental set-up and instrumentation

The instrumentation for voltammetric sensors is more complex than that for potentiometric sensors. To control the applied potential and register the current at a working electrode (and/or charge passed) a *potentiostat* is necessary—most electroanalytical voltammetric techniques are based on potential control.

Three electrodes are usually necessary (Fig. 4.1) in order to avoid passing current through the reference electrode, which would alter its potential via changes in the activities of the various species. The electrical circuit, through which the current passes, is between the working (indicator) electrode and an auxiliary electrode. The reference electrode serves in a three-electrode system to control the potential of the working electrode and thence the reactions which can occur there.

There are a few exceptions where two-electrode systems may be able to be used. First is the case of microelectrodes, where the currents are very small and so do not perturb the potential of the reference electrode; indeed, it may be possible to use a quasi-reference electrode such as a platinum wire in particular situations. Secondly is the case when physically large, low

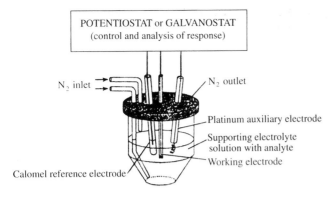

Fig. 4.1 A three-electrode electrochemical cell with control instrumentation.

impedance reference electrodes are used so that, even when currents are passed, the activities of the solution species remain essentially unaltered. This often happens in systems with dropping mercury electrodes; for example, the auxiliary electrode could be a mercury pool in contact with a halide-containing electrolyte so that the sparingly soluble salt of Hg^I is present at the surface.

The purpose of a *galvanostat* is to control the current between the working and auxiliary electrodes; the reference electrode acts as the reference voltage for the measurement of the potential of the working electrode. Thus, once again, three electrodes are, in practice, necessary (Fig. 4.1).

Control and data acquisition of the response can be conveniently done by computer through an adequate interface in a digitally based potentiostat. Analogue potentiostats and galvanostats are not now widely available, and many modern voltammetric procedures are based on step functions which lends them directly to computer control. Naturally, the digital waveform can be converted into an analogue waveform by a digital to analogue converter (DAC) and the response redigitalised through an analogue to digital converter (ADC) if necessary. Indeed, erroneous interpretation of the results can arise in linear sweep techniques if this is not done, i.e. if a digital waveform is directly applied to the electrode as a staircase waveform with too-large steps, since the theory was derived for a linear voltage ramp.

4.3 Potential sweep techniques

Potential sweep methods consist of scanning a chosen region of potential and measuring the current response arising from the electron transfer and associated reactions that occur. They are widely used for the investigation of electrode processes, which is a first step towards developing an electroanalytical procedure. Naturally, they can also give quantitative information, since the currents obtained are directly proportional to concentration.

The basic cyclic voltammetry scheme is shown in Fig. 4.2. Linear sweep voltammetry corresponds to just the first segment of the triangular waveform.

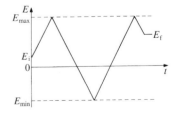

Fig. 4.2 Potential waveform for cyclic voltammetry: E_i is the initial potential, E_f the final potential, E_{max} the maximum and E_{min} the minimum potentials. The initial sweep can be in the positive or negative direction (sweep rate $v = |dE/dt|$).

A number of parameters have to be specified. Besides those depicted in the figure, the sweep rate $v = |dE/dt|$ is of extreme importance; typical values vary between 1 mV s^{-1} and 1 V s^{-1}, although much higher sweep rates are routinely used at microelectrodes.

As pointed out above, the implementation of linear sweeps in digitally based instruments is as a succession of steps in a staircase. This can lead to erroneous results, since the theory for linear sweep voltammetry and its cyclic equivalent, cyclic voltammetry (as shown in Fig. 4.2) is based on a voltage ramp and not on voltage steps. Particular care must be taken to ensure that the steps are small enough so that the two approaches are completely equivalent.

The total current obtained is a sum of a contribution from a faradaic reaction and a capacitative contribution according to

$$I = I_C + I_F = C_d \frac{dE}{dt} + I_F = vC_d + I_F \qquad (4.1)$$

I_C is proportional to v whereas, as will be shown below, I_F is proportional to $v^{1/2}$. Thus, at high scan rates, the capacitative contribution must be subtracted. In practice this is often done, as with other voltammetric techniques and usually with good precision, by subtraction of the blank response, i.e. the sweep performed under the same experimental conditions in the same solution, but without the electroactive species present.

The basic response to a linear sweep of potential is a peak-shaped curve, which can be understood as follows. The current begins to rise as potentials are reached where electrode reaction can occur. This creates a concentration gradient which sucks in more electroactive species until depletion effects set in and the current begins to fall again. Detection limits lie in the range 10^{-5} M.

In cyclic voltammetry of *reversible reactions*, i.e. those with fast electrode kinetics relative to the time-scale of the sweep, the product of the initial oxidation or reduction is then reduced or oxidised, respectively, on reversing the scan direction. Theoretical analysis of the wave shape (see Fig. 4.3a) leads to the following equation for the peak current in linear sweep voltammetry (exemplified for an oxidation)

$$I_{p,a} = 2.69 \times 10^5 n^{3/2} A D_R^{1/2} R_\infty v^{1/2} \qquad (4.2)$$

with A in cm^2, D_R in cm^2 s^{-1}, [R] in mmol cm^{-3} and v in V s^{-1}. Of particular note is the proportionality between the peak current and $v^{1/2}$. Fig. 4.3a also shows the importance of measuring the peak current of the reverse peak from a baseline which is a continuation of the decaying current response in the initial scan direction.

Other important data characteristic of reversible reactions are given in Box 4.1, including the relationship between the peak potential, E_p, and the formal electrode potential, $E^{\circ\prime}$. As the sweep rate is increased, the time-scale of the experiment becomes smaller so that, eventually, equilibrium is not reached at the electrode surface and kinetic effects begin to appear.

For completely *irreversible reactions* only the oxidation or reduction corresponding to the initial sweep direction appears, since re-reduction or re-oxidation, respectively, cannot occur, i.e. there is no reverse peak (Fig. 4.3b).

BOX 4.1 CHARACTERISTICS OF REVERSIBLE AND IRREVERSIBLE SYSTEMS $O + ne^- \rightleftharpoons R$ IN LINEAR SWEEP METHODS ILLUSTRATED FOR A REDUCTION

Reversible systems

$$I_{p,c} = -2.69 \times 10^5 n^{3/2} A D_O^{1/2} [O]_\infty v^{1/2}$$

$$E_{p,c} = E_{1/2}^r - 0.0285/n \, V$$

$$|E_{p,c} - E_{p/2,c}| = 2.2 \frac{RT}{nF} = \frac{56.6}{n} \, mV \text{ at } 298 \text{ K}$$

For cyclic voltammetry:

$$|I_{p,a}/I_{p,c}| = 1 \text{ (if scan is inverted at least 0.1 V after the peak)}$$

Reverse peak current must be measured from a baseline that is a continuation of the cathodic curve

Irreversible systems

$$E_{p,c} = E^{\ominus'} - \frac{RT}{\alpha_c n'F}\left[0.780 + \ln\frac{D_O^{1/2}}{k_0} + 0.5 \ln\left(\frac{\alpha_c n'Fv}{RT}\right)\right]$$

$$I_{p,c} = -2.99 \times 10^5 n(\alpha_c n')^{1/2} A[O]_\infty D_O^{1/2} v^{1/2}$$

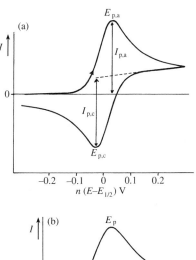

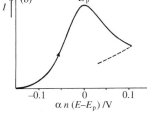

Fig. 4.3 (a) Cyclic voltammogram for a reversible reaction. (b) Linear sweep voltammogram for an irreversible reaction. (- - - -) corresponds to the continuation of current decay observed after sweep reversal.

The peak current for the initial scan is given, for an oxidation at 298 K, by

$$I_{p,a} = 2.99 \times 10^5 n(\alpha_a n')^{1/2} A R_\infty D_R^{1/2} v^{1/2} \tag{4.3}$$

where n' is the number of electrons in the rate-determining step and the units are as in Eqn 4.2. In this case the peak potential varies with the sweep rate (see Box 4.1) the effect of the kinetic limitations being to shift an oxidation to more positive potentials and a reduction to more negative potentials.

The majority of redox couples fall between the two extremes and exhibit *quasi-reversible* behaviour. This means that the reverse peak appears but is smaller than the forward peak. By measuring the peak potentials for forward and reverse scans it is possible to deduce standard rate constants from values of peak separation for simple electrode processes $O + ne^- \rightleftharpoons R$; this relation has been tabulated for a large number of values of peak separation (see Table 4.1). Such a table can be used at the same time as a quantitative determination is being carried out. Nowadays, there also exist powerful and fast computer simulation and fitting procedures which fit the whole curve and not just the peak potentials.

Adsorbed species lead to changes in the shape of the cyclic voltammogram, since they do not have to diffuse to the electrode surface. In particular, if only adsorbed species are oxidised or reduced, in the case of fast kinetics the cyclic voltammogram is symmetrical, with oxidation and reduction peak potentials coincident (Fig. 4.4).

If the adsorption energy of O and R is equal then the peak potential is equal to the formal potential of the adsorbed redox couple. As the kinetics become slower some peak separation begins to occur. Nevertheless, in all these cases the peak current is proportional to v (not $v^{1/2}$), which therefore represents a

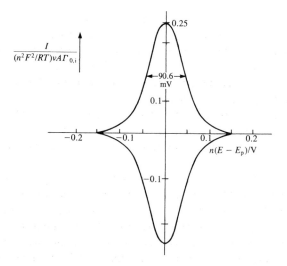

Fig. 4.4 Cyclic voltammogram for the reversible reaction of adsorbed species; $\Gamma_{R,i}$ is the initial surface concentration of adsorbed i.

Table 4.1 Relation between standard rate constants and separation of anodic and cathodic peak potentials where $\psi = k_0 v^{-1/2}(nF/\pi DRT)^{-1/2a}$

ψ	$n(E_{p,a} - E_{p,c})$/mV	ψ	$n(E_{p,a} - E_{p,c})$/mV
20	61	0.38	117
7	63	0.35	121
6	64	0.26	140
5	65	0.25	141
3	68	0.16	176
2	72	0.14	188
1	84	0.12	200
0.91	86	0.11	204
0.80	89	0.10	212
0.75	92	0.077	240
0.61	96	0.074	244
0.54	104	0.048	290
0.50	105		

[a] This assumes $D_O = D_R$ and $\alpha = 0.5$ (in practice the variation of E_p with α can be assumed not to vary in the range $0.3 < \alpha < 0.7$.

diagnostic of electrode reaction of adsorbed species, as exemplified by the oxidation peak current for a reversible process

$$I_{p,a} = \frac{n^2 F^2 v A \Gamma_{R,i}}{4RT} \tag{4.4}$$

in which $\Gamma_{R,i}$ is the surface excess of R before the sweep.

Cyclic voltammetry may also be used to investigate *multistep electrode processes* and those involving *coupled homogeneous reactions*. Simulation packages for a number of mechanisms are available. In this way, the whole voltammetric curve is fitted and optimised in order to obtain the best kinetic and thermodynamic parameters. This is a modern alternative to convolution potential sweep voltammetry which consists of the transformation of the cyclic voltammetric curve into a sigmoidal shape of exactly the same form as a steady-state voltammetric curve, to which established analysis procedures are then applied.

Finally, the question of convection–diffusion at slow sweep rates should be considered. Convection–diffusion sweeps away the products of the reaction and, if the sweep rate is sufficiently slow, there is always a steady-state current registered. If the sweep rate is increased then a steady state is not achieved and a peaked waveform begins to appear. At extremely high sweep rates the current due to the linear sweep is much higher than that from convection—the only (but very useful) function of the convection is to remove the products of the electrode reaction and to suppress natural convection.

At microelectrodes, the large induced concentration gradient from spherical diffusion means that steady-state conditions can be achieved if the sweep rate is not too high (see Section 2.9): at low scan rates a steady-state, scan rate-independent voltammogram is obtained. However, the higher mass transport, i.e. higher effective mass transfer coefficient, means that the influence of kinetics is more easily observed: reversibility is less than at larger electrodes and higher rate constants for electron transfer can be measured.

4.4   Step and pulse techniques

Many pulse techniques have been devised based on a succession of potential steps of varying height and in forward or reverse directions. They find wide application in digitally based potentiostats, which are directly suited to their exploitation. Indeed, as mentioned above, the implementation of cyclic voltammetry in many instruments is as a staircase waveform.

The basics of the current response to a potential step were alluded to in Chapter 2. The response is a pulse of current which dies away with time as the electroactive species near the electrode surface is consumed (Fig. 2.7b). Superimposed on this faradaic response is a capacitive contribution due to double layer charging which dies away more quickly, often within 50 μs (see Fig. 4.5). After this, the current is given, for a reversible system, by a modified form of the Cottrell equation, in which $I \propto t^{-1/2}$. This means that the charge $Q \propto t^{-1/2}$ which can also be used for analysis. (Note that at the DME and owing to the increasing electrode area and continuing production of fresh surface, the capacitive current diminishes, being proportional to $t^{-1/3}$ but continues throughout the drop life.)

Thus, in step and pulse techniques the current is usually sampled after the capacitive current has died away. Pulse widths are adjusted to satisfy this condition and the additional condition that time has not been allowed for natural convection effects to influence the response. However, in square wave voltammetry the capacitive contribution is eliminated by subtraction (see

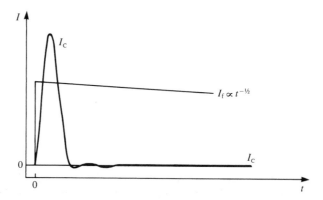

Fig. 4.5 Evolution of capacitive, I_c, and faradaic, I_F, currents after applying a potential step.

below). Experimental data are obtained in the form of a table of currents for given values of potential which are then plotted and joined together.

A number of potential pulse techniques are in common use and will now be described. Their first use was at the dropping mercury electrode, synchronising the pulse period with drop time. Since then many applications at solid electrodes and at static mercury drop electrodes have arisen which do not suffer from the difficulties associated with drop growth. Detection limits of these techniques are of the order of 10^{-7} M.

Normal pulse voltammetry (NPV)

Short potential pulses of increasing height are superimposed on a constant base potential as shown in Fig. 4.6. This base potential is chosen where no faradaic reaction occurs, i.e. where $I\,0$. Measurement of the current at the end of the pulse and plotting the succession of points against the potential of the applied pulses gives a voltammetric profile of the same form as a steady-state voltammogram, as seen in Fig. 4.6b. The maximum current is thus

$$I_{max} = \frac{nFAD^{1/2}c_{\infty}}{(\pi t_{m})^{1/2}} \tag{4.5}$$

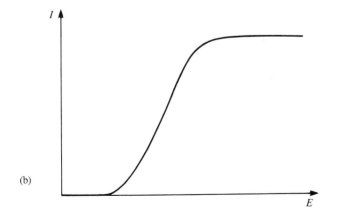

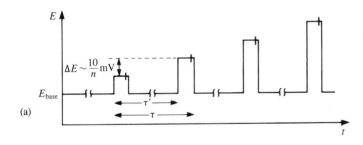

Fig. 4.6 Normal pulse voltammetry. (a) Scheme of pulse application starting at E_{base}. The current is measured at the end of the pulses and it is assumed that $I = 0$ for $E = E_{base}$. $\tau = 2-4$ s and $(\tau - \tau') = 5-100$ ms. At the DME, the end of the pulse is synchronised with drop fall. (b) Resulting $I - E$ profile.

where t_m is the sampling time after application of the pulse. An important advantage is that if irreversible adsorption of the product of the electrode reaction occurs then this will be much reduced through the use of pulses. Nevertheless, it should be remembered that a pulse leads to higher mass transport than a hydrodynamic electrode owing to the sharply induced concentration gradient, so that kinetic effects may be more visible; in particular, an electrode reaction which appears reversible at a hydrodynamic electrode may become quasi-reversible under NPV conditions. On the other hand, the short time-scale may mean that effects of coupled homogeneous reactions are not observed. Detection limits are of the order of 10^{-7} M.

A related technique is that of *reverse pulse voltammetry*. The scheme for application of potential is the same except that the base potential corresponds to diffusion-limited reaction and the pulses are applied in the reverse direction. This can be useful in situations where there are parallel electrode reactions of the initial species.

Differential pulse voltammetry (DPV)

As its name suggests, DPV measures the differences between two currents, just before the end of the pulse and just before pulse application. The base potential is incremented in a staircase as shown in Fig. 4.7. The pulse is a factor of 10 or more shorter than the period of the staircase waveform. Pulses superimposed on a potential ramp have also been employed in the past, although the staircase version is simpler to implement in digital instruments. The difference between the two sampled currents is plotted against the staircase potential and leads to a peak-shaped waveform. The peak for a reversible system occurs at a potential

$$E_p = E_{1/2} - \frac{\Delta E}{2} \tag{4.6}$$

where ΔE is the pulse amplitude (with sign included). The corresponding current, I_p, is given by the expression

$$I_p = N \frac{nFAD^{1/2}c_\infty}{(\pi t_m)^{1/2}} \left(\frac{1 - \sigma}{1 + \sigma} \right) \tag{4.7}$$

where

$$\sigma = \exp\left(\frac{nF}{RT} \frac{\Delta E}{2} \right) \tag{4.8}$$

It can be seen that, as $| \Delta E |$ increases in magnitude, the term $(1 - \sigma)/(1 + \sigma)$ also increases, becoming ± 1 for very large ΔE (reduction or oxidation, respectively), equal to the value for NPV (Eqn 4.5). However, DPV is better than NPV in the great majority of situations owing to better elimination of the contribution of non-faradaic, mainly capacitative, processes to the current signal. In NPV the charging current is larger as the pulse becomes of larger

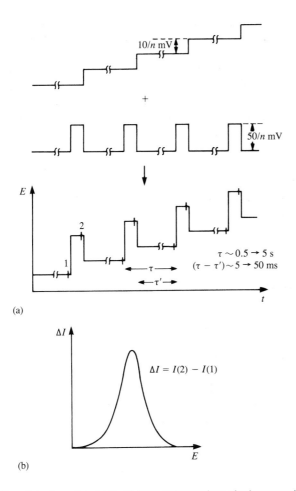

(a)

(b)

Fig. 4.7 Differential pulse voltammetry. (a) Schematic waveform of pulses superimposed on a staircase (sometimes superimposed on a ramp). (b) Schematic voltammetric profile of ΔI vs. staircase potential.

amplitude, except near the beginning of the 'scan', and, secondly, residual capacitative contributions will tend to be effectively subtracted out in DPV.

For resolving the signals due to two species with close half-wave potentials, the peak response of DPV is more useful than that of NPV. However, as the pulse amplitude increases so does the peak width, meaning that, in practice, ΔE values of more than 100 mV are not viable. The expression for the half-width at half-height, $W_{1/2}$

$$W_{1/2} = 3.52RT/nF \qquad (4.8)$$

leads to a value of 90.4 mV for $n = 1$ at 298 K showing that peaks separated by 50 mV may often be resolved. Detection limits are once again 10^{-7} M.

Square wave voltammetry (SWV)

The square wave voltammetric waveform consists of a square wave superimposed on a staircase, as demonstrated in Fig. 4.8. The currents at the end of the forward and reverse pulses are both registered as a function of staircase potential. The difference between them, the net current, is larger than either of its two component parts in the region of the peak which is centred on the half-wave potential. Capacitative contributions can be effectively discriminated against before they die away, since, over a small potential range between forward and reverse pulses, the capacity is constant and is thus annuled by subtraction. In this way the pulses can be shorter than in DPV or NPV, i.e. the square wave frequency can be higher. Instead of the effective sweep rates of 1–10 mV s^{-1} of DPV, scan rates of 1 V s^{-1} can be employed— for example, a square wave frequency of 200 Hz with 5 mV increments of potential. Detection limits of 1×10^{-8} M are achievable under optimum conditions.

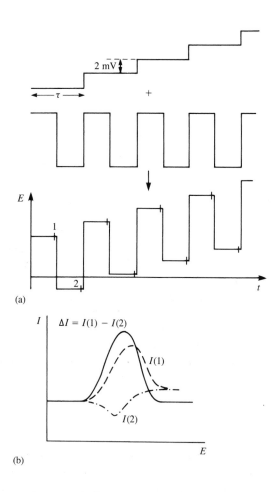

Fig. 4.8   Square wave voltammetry. (a) Schematic waveform (sum of a staircase and a square wave). (b) Schematic voltammetric profiles of *I* vs. staircase potential.

Apart from the advantages of increasing the speed of the experiment, in many electroanalytical applications in the negative potential region, oxygen does not have to be excluded from solution when using SWV unless it interferes directly with the electrode reaction under study. At rather negative potentials in the limiting current region for oxygen reduction, the forward and reverse currents are equal, leading to a zero net current. For a scan in the positive direction from negative potentials, the fast effective scan rate means that no electroactive species have time to diffuse to the electrode surface from bulk solution in a significant way. Thus, the bubbling nitrogen or argon prior to the experiment is avoided, decreasing experimental time and simplifying the procedure.

Other advantages arise from lower consumption of electroactive species relative to DPV and reduced problems of blocking the electrode surface.

Some care must be taken with the interpretation of SWV data. Whilst Fig. 4.8 shows a typical profile for the forward and reverse currents for a reversible reaction, if the electrode kinetics are slow then the forward current is shifted to higher overpotentials than the reverse current is, and the reverse current can disappear entirely. For more complex mechanisms, two peaks may appear in the net current voltammogram which, unless the individual profiles are monitored, could lead to the erroneous conclusion that two species are reacting. Simulation software packages exist which can aid in the task of unravelling mechanisms and kinetics.

Application of step and pulse techniques in electroanalysis

At present, the step and pulse techniques used most widely in electroanalysis are differential pulse voltammetry and square wave voltammetry. The latter is relatively new for such work, mainly due to former problems of electronic implementation in the instrumentation, but it is taking over a large part of the traditional domain of the application of DPV.

In conceptual terms, DPV and SWV are very similar. SWV can be thought of as a special case of a DPV waveform where the interval between pulses is equal to the pulse width, but with the important difference that the base potential, at which the current is registered on the voltammogram, is shifted in the scan direction by $|\Delta E/2|$ in DPV terms. Many instruments allow the user complete freedom to choose the waveform parameters and to cover the whole transition range from DPV, as originally conceived, to SWV by reducing the interval between DPV pulses until it becomes equal to the pulse width. It must not be forgotten that, at this limit, E_p from such a DPV scheme is shifted relative to properly specified SWV as described.

4.5   ac techniques

In ac voltammetry a small amplitude sine wave is superimposed on a succession of fixed potentials or on a linear ramp of potential. The alternating current response varies in amplitude and phase angle from the perturbation. The form of the curves obtained gives information concerning the kinetics, and the response can also be used for analytical purposes. Measuring the components of the current in-phase (0°) and out-of-phase (90°) with the applied voltage waveform enables effective separation of charging currents, which appear at 90°, and faradaic currents, which appear at 0°. The schematic response is shown in Fig. 4.9.

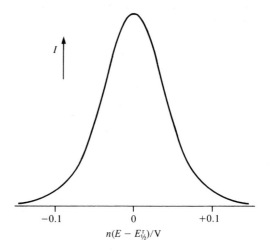

Fig. 4.9   ac voltammetry schematic response.

A reversible system gives a symmetrical peak centred on the half-wave potential and for which the peak current for an oxidation is given by

$$I_p = \frac{n^2 F^2 A \omega^{1/2} D_R^{1/2} [R]_\infty \Delta E}{4RT} \tag{4.9}$$

with ΔE the rms (root mean square) amplitude of the sine wave of frequency ω rad s^{-1}. A symmetrical wave is obtained which can be described by

$$\left(\frac{I_p}{I}\right)^{1/2} - \left(\frac{I_p - I}{I}\right)^{1/2} = \exp\left(\frac{nF(E_{dc} - E_{1/2}^r)}{2RT}\right) \tag{4.10}$$

From this the peak half-width can be deduced as 90.4/n mV at 298 K.

Detection limits of the order of 5×10^{-7} M can be attained for reversible systems. Unfortunately, slower electrode kinetics results in a substantial loss of sensitivity due to reduced peak currents, but this can be turned to advantage in solutions where one component reacts reversibly and the other irreversibly.

4.6   Increasing selectivity: membrane and modified electrodes

As can be seen from the previous sections, much selectivity can be obtained in voltammetric techniques by control of the applied potential. The potential at which the current peaks appear identifies the electroactive species. There are many situations, however, in which this is not sufficient. One reason may be the close proximity of the formal potentials of two electroactive species, or differing electrode kinetics—the result is overlapping peaks which are difficult to deconvolute. The other reason is a decrease in response with time due to blocking of the electrode surface by irreversible adsorption processes, commonly of organic compounds. The use of appropriate step and pulse techniques may help to reduce the problem of poisoning but additional measures are necessary.

Any extra selectivity has to result from modification of the electrode surface or creation of a selective physical barrrier, i.e. modified electrodes and membranes. Both approaches are currently of importance.

High specificity can often be achieved by using a biological sensing element, for example an enzyme, either immobilised on the electrode or on or within a membrane covering the electrode. These *amperometric* and *voltammetric biosensors* rely on a combination of electrochemical signal transduction together with the biological sensing component.

Membrane-covered and metallised membrane electrodes

Two approaches can be used. The first of these involves covering the electrode with a porous membrane so that the species of interest can traverse the membrane, but larger species such as proteins are unable to do so—these are known as size-exclusion membranes. They are usually in direct contact with the electrode.

Alternatively, the membrane can be separated from the electrode surface by a thin film of electrolyte. The most well-known example of such an electrode is the Clark oxygen electrode (Fig. 4.10) which allows only oxygen to traverse the membrane: oxygen either dissolved in solution or in the gas phase.

An important variant of these membranes is when the inner surface of the membrane is itself the indicator electrode—metallised membrane electrodes—which contact with an inner electrolyte solution.

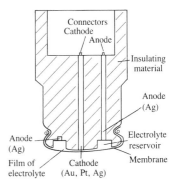

Fig. 4.10 The Clark oxygen electrode.

Modified electrodes

The essential distinction between modified electrodes and direct coverage with membranes is that surface modification involves more than just creating a physical barrier—it either changes the surface layers of the electrode itself or creates a layer with some form of chemical as well as physical selectivity. This means, in other words, that some electrode processes are enhanced whilst others are inhibited. In some cases, the modifying layer is electroactive, acting as a mediator (new electrode material) between the solution and the electrode substrate in electron transfer (Fig. 4.11).

Modified electrodes can be prepared in the following ways.

(a) *Chemical modification.* The electrode surface is activated by chemical reaction, such as with silane, which is then used to react with another chemical species that becomes immobilised on the surface. Apart from

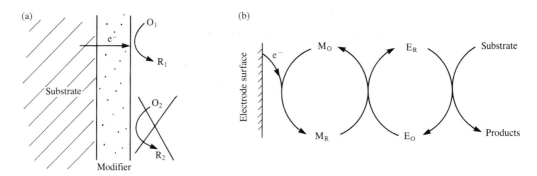

Fig. 4.11   Surface-modified electrodes showing (a) selectivity, (b) action of a mediator, M, in regenerating an enzyme, E (subscripts O and R refer to oxidised and reduced forms, respectively).

organic compounds, enzymes are often immobilised by such procedures following reaction of the surface with reagents such as carbodiimide.

(b)   *Adsorption*. This is used for coating electrode surfaces with solutions of the modifier either by dipping or, more commonly, by application of a drop of solution followed by spinning to evaporate the solvent (spin-coating). This is particularly used for modifying with soluble polymers. Some polymeric species which have a tendency for self-assembly can also be applied through such procedures, leading to self-assembled monolayers on the electrode surface.

(c)   *Electroadsorption and electrodeposition*. If adsorption is carried out under the influence of an applied potential then thicker modifier layers usually result, but there is probably a greater guarantee of uniformity. Electropolymerisation of monomers is also possible. Such procedures are used for the formation of conducting polymers. Note that in these cases, however, the electrode material is effectively that of the conducting polymer, the function of the electrode substrate being merely that of current collector.

(d)   *Plasma*. The electrode surface is cleaned by a plasma leaving the surface with dangling bonds and being highly active. Adsorption of any species, such as amines or ethenes, in the vicinity is very fast.

Bulk modification rather than surface modification procedures are sometimes used. In such cases the modifier is mixed intimately with the electrode material, such as with carbon paste. The advantage of this approach is that a new surface of similar properties can be quickly exposed by polishing or cutting.

4.7   Decreasing detection limits and increasing sensitivity: preconcentration methods

Industrial and environmental analyses are increasingly demanding lower detection limits and those indicated above are often not sufficiently low.

Table 4.2   Principles of preconcentration techniques

Method	Preconcentration step	Determination step	Measurement
SV: stripping voltammetry	Potential control	Potential control	I vs. t (or I vs. E)
AdSV: adsorptive stripping voltammetry	Adsorption (with or without applied potential)	Potential control	I vs. t (or I vs. E)
PSA: potentio-metric stripping analysis	Potential control	Reaction with oxidant, reductant in solution or applied current[a]	E vs. t

[a] In the case of applied current usually referred to as stripping chronopotentiometry.

Preconcentration of the analyte species at or on the electrode surface before the voltammetric scan, if possible, can often solve the problem and has the added advantage of simultaneously increasing sensitivity. The principles of three commonly used preconcentration techniques are shown in Table 4.2 and the responses obtained in Fig. 4.12. Collectively, these are often known as *stripping analysis*, a misnomer which arose from the first of these techniques where, after accumulation at the electrode surface, the analyte is 'stripped' into bulk solution during the determination step.

Thus the *preconcentration step* is carried out under fixed potential control, usually at a potential corresponding to the limiting current region or to maximum adsorption rate, or at open circuit. Its efficiency depends on the rate of transport of the species to be accumulated to the electrode surface; a constant rate of transport will lead to better reproducibility and repeatability and a linear dependence on accumulation time. Thus constant stirring is used with static mercury drop electrodes and stationary electrodes, as are

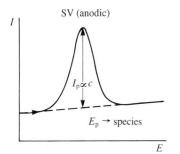
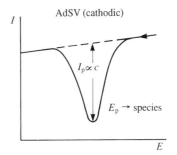
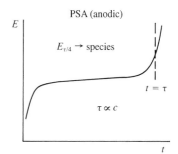

Fig. 4.12   Schematic responses in determination step of voltammetric preconcentration techniques in Table 4.2.

hydrodynamic electrodes, either rotating or in flow streams, in order to increase sensitivity and decrease detection limit.

The *determination step* is under potential control, current control or at open circuit with an oxidant or reductant in solution. Potential control, used in stripping voltammetry and adsorptive stripping voltammetry, leads to a current peak whose height (and area) is proportional to the concentration of the accumulated species and thence to the bulk solution concentration. Normally a linear potential scan, a differential pulse scan or, more frequently, a square wave scan is applied during the determination step. The advantages of the latter in terms of speed were alluded to in the previous section; also, for the analysis of heavy metal ions deposited in mercury thin film electrodes at negative potentials during preconcentration (anodic stripping voltammetry), oxygen does not need to be removed from solution. In both potentiometric stripping analysis and stripping chronopotentiometry, rather than potential control, oxidation of the accumulated species is either induced or effected by an oxidation current, and an E vs. t transient is obtained, the duration of which is proportional to concentration of the analyte. Instead of E vs. t, dt/dE vs. t is often registered, which is a peak whose height is also proportional to concentration.

Detection limits can be as low as 10^{-11} M.

Anodic stripping voltammetry (ASV)

Anodic stripping voltammetry is used for the analysis of cations in solution, mainly heavy metal cations. These are accumulated at the electrode by reduction to the zero oxidation state. Usually the electrode material is the mercury electrode, the advantage of which is to increase the negative potential range available, particularly important in the case of the commonly analysed zinc; the elements are dissolved in or form an amalgam with the mercury. The mercury electrode can either be in the form of a static or hanging mercury drop, or as a mercury thin film electrode (MTFE) on a suitable substrate such as glassy carbon, the film being either preformed or codeposited with the analyte cations. Thus, in the preconcentration step the reaction is

$$M^{n+} + ne^- \rightarrow M\ (Hg)$$

which is reversed in the determination step. The latter is now commonly accomplished by means of a square wave voltage scan, the technique being referred to as SWASV. Advantages of this are its rapidity, which means that solutions do not have to be deoxygenated since oxygen has no time to diffuse to the surface of the electrode during the determination scan and contribute to the signal (typically 1–2 s), and its high sensitivity. Important examples are the determination of copper, lead, cadmium and zinc ions at sub-ppb levels.

When mixtures of cations are analysed at mercury electrodes problems can arise due to the formation of intermetallic compounds of variable stoichiometry that are soluble in mercury. To avoid this, specific strategies have to be developed for each case but often involve the codeposition of a third element. A pertinent example is in the analysis of solutions containing both Cu^{2+} and Zn^{2+}, which can lead to the formation of Cu–Zn intermetallics in the mercury. To avoid this, the solution is spiked with Ga^{3+} and Ga is codeposited, leading to the preferential formation of Cu–Ga. In this way, Zn^{2+} can be analysed without interference, and Cu^{2+} can be deposited at more positive potentials where Zn^{2+} is not reduced in a separate experiment (see Fig. 4.13).

Non-electroactive analytes can sometimes be determined by ASV. An example is $Ru(NH_3)_6^{3+}$ which reduces Ag^+ near the surface of the electrode and which is then deposited.

Cathodic stripping voltammetry (CSV)

This is the inverse of ASV. At mercury electrodes it can be used to measure species which form sparingly soluble mercury salts such as halide ions, sulfide, cyanide, thiols and penicillins. The deposition step is, thus, schematically

$$A^{n-} + Hg \rightarrow HgA + ne^-$$

which is inverted during determination.

It is clear that CSV is not restricted to mercury electrodes. Any electrode material that, on oxidising, forms sparingly soluble salts will work, such as silver electrodes for the determination of halide ions. Another possibility is if the ion of interest oxidises to form an oxide precipitate on the electrode surface. Pb^{2+} and Mn^{2+} can be analysed in this way.

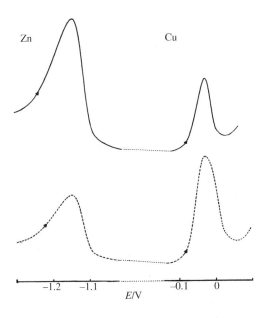

Fig. 4.13   Removal of Cu—Zn interferences in ASV of solutions containing Cu^{2+} and Zn^{2+} by spiking with Ga^{3+}: response without (- - - - -) and with (———) gallium ions.

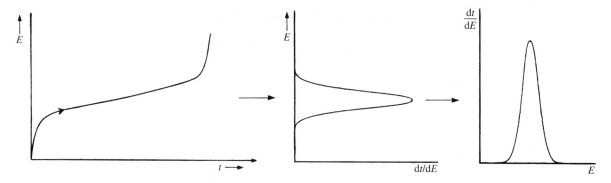

Fig. 4.14   Potentiometric stripping analysis showing transformation of *E* vs. *t* into d*t*/d*E* vs. *E* profile.

Potentiometric stripping analysis (PSA)

This technique is very similar to ASV in that the accumulation procedure is the same. The difference is that in the stripping step an oxidant in solution, or alternatively an anodic current, causes the oxidation of the reduced analyte. In mixtures the stripping potential identifies the species present and the length of time at which the potential occurs gives the concentration (see Fig. 2.8). A plot of dt/dE vs. t, which gives peaked waveforms, instead of E vs. t facilitates analysis, as shown in Fig. 4.14. The oxidant can often be dissolved oxygen so that deoxygenation is not necessary, which is an advantage over many ASV schemes except for square wave ASV (SWASV). The species which can be measured by PSA are, in general, the same as can be measured by ASV.

Adsorptive stripping voltammetry (AdSV)

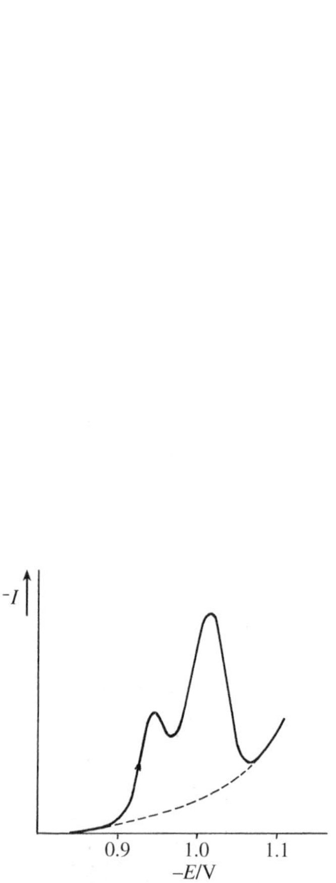

AdSV is based on the adsorption accumulation of a complex between the analyte and a complexing agent at the electrode surface, often at an applied potential chosen so as to aid the adsorption process. Thus, in principle, either the metal ion or the ligand can be measured in the presence of an excess of the other component in bulk solution. Usually the determination step involves reduction of the metal ion from the adsorbed complex. Adsorption follows an isotherm, usually of Langmuir-type, which means that an adsorption time can be reached corresponding to equilibrium saturation of the complex. In general this is avoided, it being preferred to work in the initial part of the isotherm where the adsorbed amount varies linearly with accumulation time.

The success of the AdSV strategy relies on judicious choice of a sufficiently specific chelating agent. Sometimes during the reduction determination reaction this agent can participate by recycling the partially reduced cation through reoxidation—this happens, for example, in the AdSV of Co^{2+} via formation of oxime complexes (Fig. 4.15).

Adsorbed organic compounds can sometimes be measured by making use of adsorption–desorption phenomena, even when they are not electroactive:

Fig. 4.15   Determination of Co and Ni by square wave AdSV of the oxime complexes, showing the catalytic effect on the height of the Co peak ([Ni^{2+}] = [Co^{2+}]).

$-I$

0.9   1.0   1.1
$-E$/V

desorption alters the interfacial capacity and gives rise to a tensammetric peak as the potential is scanned. The size of the tensammetric peak can be related to concentration.

When adsorption is irreversible, the electrode can be transferred from the analyte-containing solution to pure electrolyte for the determination step, in order to avoid interferences; this is known as adsorptive transfer stripping voltammetry (AdTSV).

4.8 Sensing strategies

Voltammetry is extremely versatile. Different applied potentials or applied current waveforms can be devised according to the problem to be studied. Sample preparation must also be considered; in particular, whether electrolyte has to be added to make the sample sufficiently conducting—this leads to dilution of the electroactive species and a lower signal. The strategy to be followed depends on the type of results needed, how quickly and so forth.

An important choice to be made in the elaboration of an experimental procedure or protocol is that between discrete batch analyses and continuous flow analyses. Each of these has its disadvantages. If discrete samples are available or are prepared then, at first sight, a batch analysis system may seem the most appropriate; however, depletion of reagents or build-up of products close to the electrode may influence the response. Assuming depletion becomes problematic, then the only alternative in batch mode is total electrolysis using a coulometric procedure—this does have the advantage that as long as 100% electrolysis can be guaranteed, then calibration is unnecessary. In continuous flow systems there is generally more reproducible and higher mass transport than in batch systems (assuming the solutions are stirred) owing to the high imposed convection; the disadvantage is the large consumption of sample solution. On the other hand, continuous flow systems do permit adding a derivatising reagent or reagents before the detector to make the substance electroactive, as is sometimes done in flow injection systems. With this in mind, flow injection procedures have been developed in which the sample is injected into a continuous flow carrier stream; some dilution of the sample occurs before the detector. A hybrid procedure is batch injection analysis in which the sample is injected directly over the detector. Electrolyte addition is not necessary and flow only occurs during the injection period. More details of this method can be found in Section 2.8. It should be noted that an injection of 20 μl of a solution containing electroactive species of concentration 1 nM corresponds to detection of 20 femtomoles of analyte.

For on-line monitoring, continuous flow systems are to be preferred and a number of detectors have been developed with this in mind based on the wall-jet, tube and thin-layer principles. Whilst the flow of solution is continuous, it is often possible to inject analyte samples into the flow stream, as mentioned above; the same basic scheme effectively occurs in an HPLC detector. It is clear that such continuous flow monitoring is particularly simple for fixed potential, amperometric detection, especially for discrete injections (Fig. 4.16). However, with the advent of fast scan techniques such as square wave

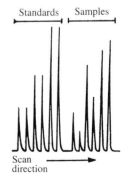

Fig. 4.16 Typical amperometric response from injection into a continuous flow stream.

voltammetry this no longer has to be the case, since the whole scan can be performed while the sample plug is passing the detector.

Another type of sensing strategy makes use of microelectrodes. Owing to the large concentration gradients which they induce, they are much more insensitive to convection than macroelectrodes, which can be a significant advantage. An important advantage can come from the fact that the current passed is so small that the ohmic drop, *IR*, is also small, even in resistive media. This means that analyses can be carried out without the necessity of high ionic concentrations, i.e. inert electrolyte does not usually have to be added, which could perturb the chemical equilibria of the system. An additional benefit of microelectrodes, apart from their small size, is that fast scan cyclic voltammetry can be done with good signal resolution owing to the much diminished capacitative contribution to the total current.

Thus, the experimental procedure to be developed will depend on the specific problem to be solved, the required sample throughput and the sample matrix. Many useful determinations need to be carried out in complex matrices, usually containing organic or biological compounds. These media encompass not only biological fluids but also many types of environmental samples such as effluents. Mineralisation removes the problem of electrode fouling but can markedly alter the speciation of the sample solution. This is strategically unsound since electrochemistry, unlike spectrometric analytical techniques, which give the total quantity of a particular element, can offer unique information on speciation and, in particular, the quantity of a given element which is labile, i.e. free to react at the electrode and not complexed or bound irreversibly to other chemical species. Thus, the aim should be, if possible, analysis of the as-received sample and protection of the electrode surface against fouling. For this purpose three strategies can be used either singly or in combination: first, use of a pulse waveform to minimise the time during which potentials (or currents) are applied that lead to irreversible adsorption; secondly, covering the electrode surface with a membrane, usually polymeric, which is permeable only to the types of species of interest, i.e. excluding and not adsorbing organic matter or modifying its surface; thirdly, reducing the contact time between sample and electrode through use of injection techniques.

5 Applications

5.1 The range of applications of electroanalysis

The objective of this chapter is to give a short overview of present developments and research in electroanalysis and its applications, together with an indication of future developments in the area. It is not possible, or indeed desirable, in a book of this kind to describe all types of modern electroanalytical procedures. Applications are particularly important in the areas of industrial, environmental and clinical analyses.

Present general challenges include the miniaturisation of electrochemical sensors, the reduction of their response time, their use on-line and automation—which includes solving all associated problems such as periodic calibration, electrode fouling and being sufficiently robust. To illustrate, examples will be chosen from potentiometric sensors, electrode surface modification, microelectrodes, flow systems, hyphenated techniques and bioelectroanalysis.

5.2 Novel potentiometric sensors

The periodic or continuous measurement of activities of chemical species and their variation with time is an extremely important area of application of electrochemical sensors, as demonstrated in Chapter 3. The most common of these is pH measurement with glass electrodes, although species such as sodium and potassium ions are measured by potentiometric sensors.

Recent tendencies have been the production of selective electrode sensors which are more robust, more reliable (i.e. needing less calibration since drift of potential is lower) and can be used in an ever wider range of situations with low maintenance, including in flowing solution, plus reduction of interferences and miniaturisation.

Some of the most important species in biological fluids are sodium, potassium and calcium cations, and chloride and carbonate anions, which cannot be measured voltammetrically. Thus, efforts have been addressed towards improved ion-selective electrodes for these species. Potentiometric sensors based on plasticised PVC membranes doped with neutral carriers have been developed for all these ions, and others. The challenge is to miniaturise the conventional electrode sensor and, in order to make it more robust, to make what is, in practice, an all-solid-state version. This is possible with polymer membranes through construction of thick film sensors on silicon substrates. Probes can be made using photolithography, thin-film metallisation, chemical etching, etc. to construct the complete sensor. It usually has micrometre dimensions, which suggests the use of multicomponent sensor arrays, although often the arrays are all fabricated to determine the same species so as to improve the signal-to-noise ratio. Some of these sensors, which are not only small but also biocompatible, can be easily sterilised and used for *in vivo* monitoring.

The other great problem with ion-selective electrodes is that of selectivity. There is much effort in improving this relative to interfering species through adequate chemical recognition principles. For example, crown and bis-crown ether ionophores can be incorporated into polymer membranes, the recognition coming from the size of the host cavity or through specific metal–ligand interactions; an alternative with similar properties is calixarenes (phenol–formaldehyde condensates) derivatised to bind metal ions. A different route is to electropolymerise conducting polymer monomers from a solution containing the counteranion which it is desired to measure, in order to tailor the cavity size in the film to that ion.

Organic compounds can also influence the response of ISEs due to interaction with the membranes. For example, surfactants are added in routine analyses to enhance washing efficiency in flow systems, besides being present in many environmental samples. Although ionic surfactants have a larger influence, as would be expected, non-ionic surfactants can also have effects due to partitioning into the polymeric membrane. Such partitioning can be used to good effect, however, for measuring the concentration of large charged organic compounds: e.g. polyions such as heparin and protamine.

ISFET sensors have had a slow but progressive evolution. Although some membrane-type sensors described above are not much larger than ISFETs, the principle of operation is conventional. The advantage of ISFETS may be the improved signal-to-noise ratio. Naturally, there is the potential to miniaturise ISFETs further to nanometer dimensions, which is probably not possible with the other sensors. Indeed, a single chip can contain multiple sensors, include signal processing circuitry and thus lead to mass fabrication of disposable sensors.

5.3 Surface-modified voltammetric electrode sensors

Surface modification can be done to increase electrode selectivity or to prevent fouling. There are many examples of such sensors. Fouling can arise from interfering species or from the analyte itself or its electrode reaction products, as happens with pesticides and herbicides, for example. In this section three types of modification will be chosen as examples.

The first involves modification by monolayer films. This can be done in the conventional way by chemical reaction of the surface. Alternatively, electrodes can be coated by self-assembled monolayers (SAMs). For example, gold electrodes may be modified by *n*-alkane thiols simply by placing the clean gold electrode in a solution containing the desired alkanethiol—this leads to a well-organised, densely packed monolayer with the alkane chains arranged in a perpendicular fashion relative to the electrode. These are selective towards hydrophobic compounds such as noradrenaline and dopamine and discriminate against hydrophilic compounds such as ascorbic acid. Such SAMs are simpler to make than monolayers by the Langmuir–Blodgett technique.

A second type involves covering the electrode with a polymer layer. Apart from physical coating, conducting polymer layers may be grown by electropolymerisation on the surface. Various types of polymer are currently under investigation, including polymer mixtures; many have some cation or anion selectivity and can be referred to as ion-exchange polymers—an example of the former is Nafion due to its sulfonate groups. Most of them have the additional advantage that protein and surfactant adsorption is much reduced relative to the bare electrode material. Much technological effort has been devoted to making thin robust films. Sol–gel polymers in which an ionomer is entrapped with a rigid gel matrix are another possibility: for example, poly(vinylsulfonic acid) or chemically modified graphite. Inorganic polymer layers, namely clays and zeolites, are also of current interest as these show different adsorption properties towards potential analytes. Note that polymer layers can be made selective to gases in solution and also in the gas phase.

A third type of modified electrode involves bulk modification. Sol–gel polymers are, in effect, an example of these. Enzymes may be immobilised within polymers by mixing. Modified carbon paste electrodes may be produced by mixing with a polymer, biological compound, enzyme or tissue; the surface of porous carbon composite electrodes can be modified.

5.4 Microelectrodes and microvolumes

The exceptional increase in the use of microelectrodes over the last two decades has had many reflections in electroanalysis. The necessity of addition of electrolyte to samples can often be bypassed; the problems of contamination are lessened. Miniaturisation, together with multicomponent detection, becomes much easier. At present, microelectrodes with submicrometre dimensions are starting to be fabricated; if these are to be used singly, the instrumentation becomes of primary importance since the currents registered are at the picoampere level. The use of assemblies of microelectrodes all with the same function can increase the total measured current whilst retaining the particular properties of microelectrodes with respect to high concentration gradients. Microelectrodes can be employed to analyse small or large volumes of solution. In the latter case one of the big advantages can be the fact that they are very insensitive to convection. Since microelectrode are local probes, they can scan across the surface, operating as potentiometric or amperometric/voltammetric sensors, and map concentrations of species—such information can be extremely valuable in the design and optimisation of robust and reliable sensors.

Other interesting developments include the microelectrode analogue of hydrodynamic double electrodes (Section 2.9). These can be made, for example, as interdigitated microband electrodes with alternate microbands connected together or as ring-disc electrodes. Some thick layer printed electrodes also have microelectrode array properties.

5.5   Continuous flow and injection methods

There is no doubt that measurements in flow systems—continuous flow and injection—are an extremely important area of electroanalysis and detectors and instrumentation have been developed with this in mind. Flow permits the easy use of multicomponent detectors, either in parallel or sequentially. It is only in this way that the ideal of real-time analysis can be approached in on-line monitoring systems. Detectors either can be placed directly in the flow streams, or samples can be removed and injected into a carrier stream or over the detector. The result of such measurements is always information on the labile species. Usually it is this that is required but even if not, in many situations it is preferable to have a fast answer, even if it is not totally accurate, than having to wait for sample digestion.

Particular advantages of continuous flow methods arise, first from the fact that fresh solution is constantly being transported to the electrode surface so that there is no reagent depletion and, secondly, because there is no build-up of products since they are continuously washed away. In flow streams for on-line monitoring the fouling of the sensor can be a big problem since there is continuous contact between the electrode and fresh contaminant. Apart from the chemical modification strategies alluded to above, the flow stream can be made to pass over a microdialysis membrane which does not allow large molecules, such as proteins, to pass, and the detector is placed behind.

The disadvantage of continuous flow analysis systems can be the high consumption of electrolyte solution that has to be added to the sample if this is not sufficiently conducting. In such cases an alternative approach has to be adopted based on a discrete analysis, i.e. injection. A common advantage of injection techniques is that if the sample is a complex matrix, such as an effluent or a biological fluid, then the degree of contamination of the electrode surface is very small. The injection (typically 20–100 μl) is either done into an inert electrolyte carrier stream (flow injection) or directly over the centre of the detector (batch injection analysis). The advantage of flow injection is that reagents can be added at any point before detection, but some dispersion of the sample plug will occur, which may be necessary for it to conduct. The advantage of batch injection analysis is its simplicity and, if the detector is surrounded by electrolyte, the sample (which arrives perpendicularly at the detector as a fine jet) can be measured directly without electrolyte addition; sample dispersion is essentially zero. Trace heavy metals and other species in solution can be measured in this way.

Flow techniques are particularly important, as is the use of microelectrodes, for the determination of environmentally toxic species such as pesticides and heavy metals, usually by stripping analysis. A particular challenge is extending such procedures to real-time analyses of samples in hazardous waste sites and effluents.

5.6 Hyphenated techniques

Hyphenated techniques involve the combination of two analytical techniques. The first part involves some sort of separation and the second the determination. There now exist many methods for the separation of components in mixtures, together with a variety of detectors. Amongst those of particular current interest are high pressure liquid chromatography (HPLC), including ion chromatography and supercritical fluid chromatography, and capillary electrophoresis. Separation under high pressure can be enhanced by the use of supercritical fluids such as carbon dioxide which allows whole samples from industrial and environmental matrices to be introduced into the column without any pretreatment to prevent column deactivation. Ion chromatography can now be used for separating heavy metal ions, which can give extra selectivity in stripping analysis.

Electrochemical detectors can provide extra selectivity compared with, for example, UV–vis detectors; they can be amperometric/voltammetric, conductometric, coulometric or potentiometric. The type of flow-through detector is usually impinging jet, tubular or thin-layer. Unfortunately, often in routine analysis, insufficient attention is paid to the problems of electrode fouling and the necessity to perform periodic routine calibrations.

An alternative to chromatographic separation is capillary electrophoresis, which is highly efficient and fast, and has a low operational cost. The capillaries have typical diameters in the region of 2 to 200 μm and lengths of 10 to 100 cm, corresponding to a volume in the picolitre to nanolitre range. The open-ended capillary, filled with the sample solution, links two buffer solutions. Voltages of the order of 10 – 30 kV are applied across the ends of the capillary which leads to electrophoretic separation of the components. The detector has to be placed directly at one end of the capillary and is essentially a modified HPLC detector. It can be readily understood that the main difficulty is the exact positioning of the detector relative to the capillary. This can be solved by using an electrode much larger than the capillary diameter placed directly against the end of the capillary, by inserting a carbon fibre microelectrode into the end of the capillary or, alternatively, by depositing a metal film, which acts as an electrode, directly on to the capillary tip.

Other hyphenated techniques use electrochemistry to preconcentrate the analyte of interest through electrolysis. This is then stripped from the electrode and aspirated into the chamber of a spectrometer for analysis: for example, inductively coupled plasma with atomic emission or mass spectrometry or an electrospray mass spectrometer. In this way the speciation advantages of electrochemistry are exploited.

5.7   Bioelectroanalysis

Bioelectroanalysis is the application of electroanalysis to the determination of biological compounds.

Many current developments in electroanalysis involve *biosensors*. This is not only because of the many advances in the life sciences that have occurred as a result of improvements in instrumentation and techniques of many kinds, but also because many important analytes are organic or biological and much attention is paid to tailoring biosensors to specific needs, making use of their natural selectivity and specificity. In particular, a lot of effort has been devoted to glucose sensors owing to their exceptional importance for diabetics. A biosensor contains a biological sensing element that responds to the analyte, such as an enzyme, which is converted by the transducer into an electrical signal. A mediator is often placed between the sensing element and the electrode substrate to enable electron flow to or from the substrate. A problem associated with biological sensing elements is that they tend to have less stability than their inorganic analogues.

Potentiometric and voltammetric biosensors have been influenced by all the trends described in previous sections: surface modification, miniaturisation and use in flow systems.

One of the main goals is to perform *in vivo* electrochemistry, and efforts have been devoted to electrode implantation in the brain and in other tissues. One problem that remains is biocompatibility and, even when this is achieved, the natural reaction of the invaded tissue is to protect itself from the invading microelectrode.

Bibliography

Literature references to specific investigations or novel contributions will not be made here. Instead, a bibliography of journals that publish many of the exciting developments in electroanalysis is given, which the interested reader should consult.

Analyst
Analytica Chimica Acta
Analytical Chemistry
Analytical Letters
Bioelectrochemistry
Biosensors and Bioelectronics
Electroanalysis
Electrochimica Acta

Fresenius' Journal of Analytical Chemistry
Journal of Electroanalytical and Interfacial Electrochemistry
Journal of Solid State Electrochemistry
Sensors and Actuators
Talanta

Other chemistry journals also include articles on electroanalysis.

Appendix 1. Data analysis

A1.1 The importance of data analysis

Data obtained in any experiment have to be subjected to widely accepted, stringent analysis procedures to assess their quality, in particular their *accuracy*, i.e. closeness to the truth, and their *precision*, i.e. reproducibility. Thus, the importance of objective analysis of data obtained in any analytical procedure cannot be overstressed. For example, new electroanalytical procedures are often subjected to method validation in which comparison is made with another technique, such as electrothermal atomisation spectroscopy, or interlaboratory comparisons are done where the same samples are analysed in different laboratories by the same procedure.

Criteria that can affect the quality of the data may be linked to the cost of performing the analyses and how competent the technical operator needs to be. With present trends in the computer control of analytical experiments, including data analysis, chemical analysis is becoming more user friendly and the level of expertise necessary in order to obtain meaningful results is becoming less high. The other side of the coin is that apparently excellent, but wrong, results can be obtained by operators without sufficient background knowledge to be able to interpret their raw results correctly, know the limitations of the procedures, and understand when the information obtained is not meaningful, and the reasons why not, in order to be able to modify the experimental procedure in a satisfactory way.

The objective of this appendix is to give a brief summary of some of the commonly used methods for evaluation of electroanalytical data. For a more exhaustive and more detailed approach the reader should consult some of the texts listed at the end of the appendix.

A1.2 Mean and standard deviation

Nearly all data analytical methods are based, in one way or another, on the *mean*, x, and *standard deviation*, s, of the distribution of n experimental results according to

$$\bar{x} = \frac{\sum_i x_i}{n} \tag{A1.1}$$

$$s = \sqrt{\frac{\sum_i (x_i - \bar{x})^2}{(n-1)}} \tag{A1.2}$$

These equations assume that errors can be equally positive or negative about the mean value. If this is the case, i.e. there are no systematic errors, then x is equal to the mean of the *normal* or *Gaussian distribution*. In the true Gaussian distribution (Fig. A1.1) with mean μ and standard deviation σ (the same equation as A1.2 but with n in the denominator), 68% of the values lie between $\mu - \sigma$ and $\mu + \sigma$, 95% between $\mu - 2\sigma$ and $\mu + 2\sigma$ and 99.7% between $\mu - 3\sigma$ and $\mu + 3\sigma$. This is the basis of confidence intervals and confidence limits, which will be discussed below. Another important quantity is the *standard error of the mean* (sem)

$$\text{sem} = \frac{\sigma}{\sqrt{n}} \tag{A1.3}$$

The sem is a measure of the uncertainty involved in estimating μ from the experimental mean x, reflecting the fact that the more measurements are made the greater the reliablity in measuring μ.

A1.3 Evaluation of the quality of sets of data

In developing a new analytical procedure, comparison often has to be made with an accepted 'accurate' value or with another procedure, possibly already established. There are two important data analysis methods for such comparisons, the *t-test* and the *F-test*. However, before applying them each data set should be tested for outliers, i.e. values that appear to be unreasonably distant from the others in the set; the objective way of doing this is by using the *Q-test*.

The Q-test

The rejection of outliers from a set of data by the *Q*-test involves calculating

$$Q = \left| \frac{\text{suspect value} - \text{nearest value}}{\text{max.value} - \text{min.value}} \right| \tag{A1.4}$$

from the experimental values and comparing with tabulated critical values of Q. Critical values for confidence levels of 90 and 95% are shown in Table A1.1.

Table A1.1 Critical values of Q for rejection of data

Number of observations	Q_{crit}	
	90% confidence	95% confidence
4	0.76	0.83
5	0.64	0.72
6	0.56	0.62
7	0.51	0.57
8	0.47	0.53
9	0.44	0.49
10	0.41	0.47

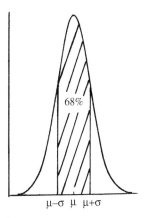

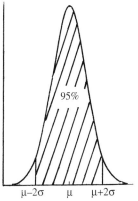

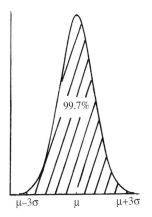

Fig. A1.1 The Gaussian distribution showing the percentage of values lying between $\mu - n\sigma$ and $\mu + n\sigma$.

If the calculated Q exceeds the critical value then the suspect value is rejected. Clearly, the suspect value must be either the maximum value or the minimum value. If a value is rejected then the test should be repeated with the smaller data set until no further data are removed.

The t-test

The *t*-test was developed for comparing means and testing for systematic errors, even though precision is also taken into account.

Comparison between an accepted value and an experimental data set. For a Gaussian distribution of data, the percentage of the distribution lying between specified limits around the mean value can be calculated. This is the width of the confidence interval corresponding to that percentage for an infinite population (infinite number of degrees of freedom).

Analogously, for a finite number of analyses, the equation describing the limits within which the true mean must lie at a given confidence level with respect to the experimental mean, *x*, is

$$\mu = \bar{x} \pm \frac{ts}{\sqrt{n}} \tag{A1.5}$$

Thus, for there to be no significant difference

$$|\mu - \bar{x}| \geq \frac{ts}{\sqrt{n}} \tag{A1.6}$$

Values of t are tabulated according to the number of degrees of freedom, in this case $(n - 1)$, and depend on the number of analyses performed and the confidence interval, i.e. what percentage of the hypothetical Gaussian distribution is to be included. Confidence levels are also presented as probabilities, P, of differences being found, e.g. a confidence level of 95% corresponds to $P - 0.05$. Some of these values are shown in Table A1.2. Note that an infinite number of degrees of freedom corresponds exactly to the Gaussian distribution.

Occasionally, a bias of experimental results in one particular direction may be suspected, equivalent to a systematic error. The *t*-test can be employed to evaluate whether the bias is significant using a so-called one-sided (one-tailed) test, instead of the normal two-sided test. The equations are the same but the *P*-values, such as those in Table A1.2, should be halved and the corresponding confidence levels altered accordingly.

Comparison of two experimental data sets. The degree of overlap of the two distributions is evaluated, i.e. if there are significant differences between the data series at a specified level of confidence. Expression A1.6 now becomes

$$|\mu - \bar{x}| \geq ts \sqrt{\frac{n_1 + n_2}{n_1 n_2}} \tag{A1.7}$$

Table A1.2 The *t*-distribution: values of *t*

Confidence interval	90%	95%	98%	99%
Significance level, *P*	0.10	0.05	0.02	0.01

Degrees of freedom				
1	6.31	12.71	31.82	63.66
2	2.92	4.30	6.96	9.92
3	2.35	3.18	4.54	5.84
4	2.13	2.78	3.75	4.60
5	2.02	2.57	3.36	4.03
6	1.94	2.45	3.14	3.71
7	1.89	2.36	3.00	3.50
8	1.86	2.31	2.90	3.36
9	1.83	2.26	2.82	3.25
10	1.81	2.23	2.76	3.17
12	1.78	2.18	2.68	3.05
14	1.76	2.14	2.62	2.98
16	1.75	2.12	2.55	2.92
18	1.73	2.10	2.55	2.88
20	1.72	2.09	2.53	2.85
30	1.70	2.04	2.46	2.75
∞	1.64	1.96	2.33	2.58

where s is the pooled standard deviation from the two sets of data, given by

$$s = \sqrt{\frac{(n_1 - 1)s_1^2 + (n_2 - 1)s_2^2}{n_1 + n_2 - 2}} \qquad (A1.8)$$

A potentially useful application of this formula is in the prediction of *detection limits*. One series of data represents already measured blank values, of number n_b, and the other the data to be measured, of number n_1. If the standard deviation is assumed to be equal for both series then the minimum number of measurements necessary to attain a predetermined detection limit, Δx_{min}, can be calculated for a given confidence level through application of

$$\Delta x_{min} = ts_b \sqrt{\frac{n_1 + n_b}{n_1 n_b}} \qquad (A1.9)$$

The F-test

The *F*-test compares the precision of two sets of data via the expression

$$F = \frac{s_1^2}{s_2^2} \qquad (A1.10)$$

where $s_1 > s_2$. If the calculated F is greater than a critical value that is tabulated for the chosen confidence level, then there is a significant difference at that probability level. Some values are shown in Table A1.3, where v_1 and v_2 are the number of degrees of freedom of each data set.

Table A1.3   Values of F at the 95% confidence level ($P = 0.05$)

v_2	v_1							
	2	3	4	5	6	7	8	∞
2	19.00	19.16	19.25	19.30	19.33	19.35	19.37	19.50
3	9.55	9.28	9.12	9.01	8.94	8.89	8.85	8.53
4	6.94	6.59	6.39	6.26	6.16	6.09	6.04	5.63
5	5.79	5.41	5.19	5.05	4.95	4.88	4.82	4.36
6	5.14	4.76	4.53	4.39	4.28	4.21	4.15	3.67
7	4.74	4.35	4.12	3.97	3.87	3.79	3.73	3.23
8	4.46	4.07	3.84	3.69	3.58	3.50	3.44	2.93
∞	3.00	2.60	2.37	2.21	2.10	2.01	1.94	1.00

A1.4   Calibration and standard addition plots—detection limits

Calibration plots are extremely important in quantitative analysis. Usually, a linear relationship between concentration and response is required. Calibration plots can be obtained by constructing plots of response vs. standard concentrations encompassing the concentration range of interest; the unknown is then analysed by interpolation. Alternatively, once the adequacy of the procedure has been proved, the *standard addition method* can be employed. In this method, after the sample solution of unknown concentration is analysed, it is spiked several times with a solution of a standard in such a way so as to give equally spaced increasing concentrations of the species to be determined. A plot of response vs. concentration of standard is constructed; the concentration of the unknown can be found from the intercept of the plot on the concentration axis.

In either of these cases, and assuming that a linear calibration plot is obtained, it is important to determine the line of best fit and the respective error. Known concentrations are plotted on the x-axis and the response on the y-axis, and it is assumed that all errors are in the y-values. It is also generally assumed that the magnitude of errors in y is independent of analyte concentration, although the statistical analysis can be modified to take varying error magnitudes into account.

The first indicator of a good straight line is the correlation coefficient, r, varying between -1 (perfect correlation on a line of negative slope) and $+1$ (perfect correlation on a line of positive slope). In practice, numerical values of at least 0.99 are necessary for what would be referred to visually as a reasonable straight line.

The next step is to calculate the line of best fit, which is usually done by minimising the sum of the squares of the residuals from the line of best fit (method of least squares) leading to the slope and the intercept. Most pocket calculators will do this calculation as well as calculating the correlation coefficient. Usually, they do not calculate the errors in the slope and intercept directly, although it is common for computer statistics software packages to do so.

The detection limit is defined as the minimum value of the signal from the species being measured that is significantly different from the blank signal. Although there is some discussion on this point, the recent tendency is to base its calculation from calibration plots on three times the standard deviation of the y-residuals. Thus, the intercept of the linear plot, $y = a + bx$, is the blank value of the response y_B ($= a$), and the limit of detection (lod) for x corresponds to the y-value for $y = y_B + 3s_B$, where s_B is the standard deviation of the y-residuals from the line of best fit. Once again many software packages provide the value of s_B directly. The limit of detection is then ($3s_B/b$) as shown in Fig. A1.2.

Some thought concerning this method for determining detection limits reveals a potential weakness which can be of great importance. If the calibration plot is of high quality, then s_B will be very small, leading to an extremely low detection limit. In practice, however, such a limit may not be attainable owing to limitations of the experimental procedure or the instrumentation (noise and drift at low signal levels), apart from any chemical interferences. Practical detection limits determined experimentally, corresponding to when it is not possible to distinguish the signal from the background visually, are of the greatest importance. Fortunately, in many cases the difference between the two approaches is not large.

More complicated statistical analyses, including non-linear regression, can be treated. The reader is referred to the bibliography at the end of the appendix for details.

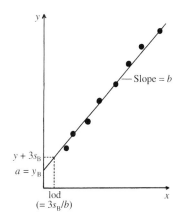

Fig. A1.2   Determination of detection limit from a linear calibration plot.

A1.5   Experimental design and factorial analysis

Recently, more elegant ways of analysing complex experimental data have been developed. These are based on consideration and identification of all the factors that can influence the outcome of the experiment and using statistical analysis to separate their influences; this is known as *factorial analysis*. The analysis of variance (ANOVA) can be employed for this purpose for both random-effect factors and factors that show a trend over a period of time. In the latter case the order of the experimental results is randomised so as to permit the use of ANOVA techniques. The exploitation of this type of analysis is referred to as *chemometrics*.

The capacity to separate the contributing factors leads to sophisticated experimental design patterns that permit the minimisation of the number of experiments necessary to characterise a system completely, since several influencing parameters can be varied simultaneously. Additionally, the application of pattern recognition techniques to the large masses of data obtained in automated experiments may allow relationships to be established that would otherwise be hidden.

Bibliography

E. Morgan, *Chemometrics: experimental design*, Wiley, Chichester, 1991.

J.C. Miller and J.N. Miller, *Statistics for analytical chemistry*, 2nd edn, Ellis Horwood, Chichester, 1992.

L. Davies, *Efficiency in research, development and production: the statistical design and analysis of chemical experiments*, RSC, Cambridge, 1993.

Appendix 2. Standard electrode potentials

There follows a list of standard electrode potentials for common electrochemical reactions in aqueous solution, i.e. measured against the standard hydrogen electrode at 25°C (298.15 K) with all species at unit activity. Most of the values were taken from A.J. Bard, R. Parsons and J. Jordan, *Standard potentials in aqueous solution* (eds), Dekker, New York, 1985, where values for many other electrochemical reactions may also be found.

	$E^{\ominus}/\text{V}$
$Ag^+ + e^- \rightleftharpoons Ag$	+0.80
$Ag^{2+} + e^- \rightleftharpoons Ag^+$	+1.98
$AgBr + e^- \rightleftharpoons Ag + Br^-$	+0.07
$AgCl + e^- \rightleftharpoons Ag + Cl^-$	+0.22
$AgI + e^- \rightleftharpoons Ag + I^-$	−0.15
$Al^{3+} + 3e^- \rightleftharpoons Al$	−1.68
$As + 3H^+ + 3e^- \rightleftharpoons AsH_3$	−0.23
$As(OH)_3 + 3H^+ + 3e^- \rightleftharpoons As + 3H_2O$	+0.24
$AsO(OH)_3 + 2H^+ + 2e^- \rightleftharpoons As(OH)_3 + H_2O$	+0.56
$Au^+ + e^- \rightleftharpoons Au$	+1.83
$Au^{3+} + 3e^- \rightleftharpoons Au$	+1.52
$Ba^{2+} + 2e^- \rightleftharpoons Ba$	−2.92
$Be^{2+} + 2e^- \rightleftharpoons Be$	−1.97
$Br_2(l) + 2e^- \rightleftharpoons 2Br^-$	+1.06
$Br_2(aq) + 2e^- \rightleftharpoons 2Br^-$	+1.09
$BrO^- + H_2O + 2e^- \rightleftharpoons Br^- + 2OH^-$	+0.76
$2HOBr + 2H^+ + 2e^- \rightleftharpoons Br_2 + 2H_2O$	+1.60
$2BrO_3^- + 12H^+ + 10e^- \rightleftharpoons Br_2 + 6H_2O$	+1.48
$BrO_4^- + 2H^+ + 2e^- \rightleftharpoons BrO_3^- + H_2O$	+1.85
$CO_2 + 2H^+ + 2e^- \rightleftharpoons CO + H_2O$	−0.11
$CO_2 + 2H^+ + 2e^- \rightleftharpoons HCOOH$	−0.20
$2CO_2 + 2H^+ + 2e^- \rightleftharpoons H_2C_2O_4$	−0.48
$Ca^{2+} + 2e^- \rightleftharpoons Ca$	−2.84
$Cd(OH)_2 + 2e^- \rightleftharpoons Cd + 2OH^-$	−0.82
$Cd^{2+} + 2e^- \rightleftharpoons Cd$	−0.40
$Ce^{3+} + 3e^- \rightleftharpoons Ce$	−2.34
$Ce^{4+} + e^- \rightleftharpoons Ce^{3+}$	+1.72
$Cl_2 + 2e^- \rightleftharpoons 2Cl^-$	+1.36

	$E^{\ominus}/V$
$ClO^- + H_2O + 2e^- \rightleftharpoons Cl^- + 2OH^-$	$+0.89$
$2HOCl + 2H^+ + 2e^- \rightleftharpoons Cl_2 + 2H_2O$	$+1.63$
$HClO_2 + 2H^+ + 2e^- \rightleftharpoons HOCl + H_2O$	$+1.68$
$ClO_3^- + 3H^+ + 2e^- \rightleftharpoons HClO_2 + H_2O$	$+1.18$
$ClO_3^- + 2H^+ + e^- \rightleftharpoons ClO_2 + H_2O$	$+1.17$
$ClO_4^- + 2H^+ + 2e^- \rightleftharpoons ClO_3^- + H_2O$	$+1.20$
$Co^{2+} + 2e^- \rightleftharpoons Co$	-0.28
$Co^{3+} + e^- \rightleftharpoons Co^{2+}$	$+1.92$
$Co(NH_3)_6^{3+} + e^- \rightleftharpoons Co(NH_3)_6^{2+}$	$+0.06$
$Co(phen)_3^{3+} + e^- \rightleftharpoons Co(phen)_3^{2+}$	$+0.33$
$Co(C_2O_4)_3^{3-} + e^- \rightleftharpoons Co(C_2O_4)_3^{4-}$	$+0.57$
$Cr^{2+} + 2e^- \rightleftharpoons Cr$	-0.90
$Cr_2O_7^{2-} + 14H^+ + 6e^- \rightleftharpoons 2Cr^{3+} + 7H_2O$	$+1.38$
$Cr^{3+} + 3e^- \rightleftharpoons Cr$	-0.74
$Cs^+ + e^- \rightleftharpoons Cs$	-2.92
$Cu^+ + e^- \rightleftharpoons Cu$	$+0.52$
$Cu^{2+} + 2e^- \rightleftharpoons Cu$	$+0.34$
$Cu^{2+} + e^- \rightleftharpoons Cu^+$	$+0.16$
$CuCl + e^- \rightleftharpoons Cu + Cl^-$	$+0.12$
$Cu(NH_3)_4^{2+} + 2_e^- \rightleftharpoons Cu + 4NH_3$	-0.00
$F_2 + 2e^- \rightleftharpoons 2F^-$	$+2.87$
$Fe^{2+} + 2e^- \rightleftharpoons Fe$	-0.44
$Fe^{3+} + 3e^- \rightleftharpoons Fe$	-0.04
$Fe^{3+} + e^- \rightleftharpoons Fe^{2+}$	$+0.77$
$Fe(phen)_3^{3+} + e^- \rightleftharpoons Fe(phen)_3^{2+}$	$+1.13$
$Fe(CN)_6^{3-} + e^- \rightleftharpoons Fe(CN)_6^{4-}$	$+0.36$
$Fe(CN)_6^{4-} + 2e^- \rightleftharpoons Fe + 6CN^-$	-1.16
$2H^+ + 2e^- \rightleftharpoons H_2$	0 (by definition)
$2H_2O + 2e^- \rightleftharpoons H_2 + 2OH^-$	-0.83
$H_2O_2 + H^+ + e^- \rightleftharpoons HO^{\cdot} + H_2O$	$+0.71$
$H_2O_2 + 2H^+ + 2e^- \rightleftharpoons 2H_2O$	$+1.76$
$Hg^{2+} + 2e^- \rightleftharpoons 2Hg$	$+0.80$
$Hg_2Cl_2 + 2e^- \rightleftharpoons 2Hg + 2Cl^-$	$+0.27$
$Hg^{2+} + 2e^- \rightleftharpoons Hg$	$+0.86$
$2Hg^{2+} + 2e^- \rightleftharpoons Hg_2^{2+}$	$+0.91$
$Hg_2SO_4 + 2e^- \rightleftharpoons 2Hg + SO_4^{2-}$	$+0.62$
$I_2 + 2e^- \rightleftharpoons 2I^-$	$+0.54$
$I_3^- + 2e^- \rightleftharpoons 3I^-$	$+0.53$
$2HOI + 2H^+ + 2e^- \rightleftharpoons I_2 + 2H_2O$	$+1.44$
$2IO_3^- + 12H^+ + 10e^- \rightleftharpoons I_2 + 6H_2O$	$+1.20$
$IO(OH)_5 + H^+ + e^- \rightleftharpoons IO_3^- + 3H_2O$	$+1.60$
$In^+ + e^- \rightleftharpoons In$	-0.13
$In^{3+} + 2e^- \rightleftharpoons In^+$	-0.44
$In^{3+} + 3e^- \rightleftharpoons In$	-0.34
$K^+ + e^- \rightleftharpoons K$	-2.93

	$E^{\ominus}/\text{V}$
$Li^+ + e^- \rightleftharpoons Li$	-3.04
$Mg^{2+} + 2e \rightleftharpoons Mg$	-2.36
$Mn^{2+} + 2e^- \rightleftharpoons Mn$	-1.18
$Mn^{3+} + e^- \rightleftharpoons Mn^{2+}$	$+1.51$
$MnO_2 + 4H^+ + 2e^- \rightleftharpoons Mn^{2+} + 2H_2O$	$+1.23$
$MnO_4^- + 8H^+ + 5e^- \rightleftharpoons Mn^{2+} + 4H_2O$	$+1.51$
$MnO_4^- + e^- \rightleftharpoons MnO_4^{2-}$	$+0.56$
$MoO_4^{2-} + 4H_2O + 6e^- \rightleftharpoons Mo + 8OH^-$	-0.91
$NO_3^- + 2H^+ + e^- \rightleftharpoons NO_2 + H_2O$	$+0.80$
$NO_3^- + 4H^+ + 3e^- \rightleftharpoons NO + 2H_2O$	$+0.96$
$NO_3^- + H_2O + 2e^- \rightleftharpoons NO_{2^-} + 2OH^-$	$+0.01$
$Na^+ + e^- \rightleftharpoons Na$	-2.71
$Ni^{2+} + 2e^- \rightleftharpoons Ni$	-0.26
$Ni(OH)_2 + 2e^- \rightleftharpoons Ni + 2OH^-$	-0.72
$NiO_2 + 2e^- \rightleftharpoons Ni^{2+} + 2H_2O$	$+1.59$
$O_2 + 2H_2O + 4e^- \rightleftharpoons 4OH^-$	$+0.40$
$O_2 + 4H^+ + 4e^- \rightleftharpoons 2H_2O$	$+1.23$
$O_2 + e^- \rightleftharpoons O_2^-$	-0.33
$O_2 + H_2O + 2e^- \rightleftharpoons HO_2^- + OH^-$	-0.08
$O_2 + H^+ + e^- \rightleftharpoons HO_2$	-0.13
$O_2 + 2H^+ + 2e^- \rightleftharpoons H_2O_2$	$+0.70$
$P + 3H^+ + 3e^- \rightleftharpoons PH_3$	-0.06
$HPO(OH)_2 + 3H^+ + 3e^- \rightleftharpoons P + 3H_2O$	-0.50
$HPO(OH)_2 + 2H^+ + 2e^- \rightleftharpoons H_2PO(OH) + H_2O$	-0.50
$PO(OH)_3 + 2H^+ + 2e^- \rightleftharpoons HPO(OH)_2 + H_2O$	-0.28
$Pb^{2+} + 2e^- \rightleftharpoons Pb$	-0.13
$PbO_2 + 4H^+ + 2e^- \rightleftharpoons Pb^{2+} + 2H_2O$	$+1.70$
$PbSO_4 + 2e^- \rightleftharpoons Pb + SO_4^{2-}$	-0.36
$Pt^{2+} + 2e^- \rightleftharpoons Pt$	$+1.19$
$Rb^+ + e^- \rightleftharpoons Rb$	-2.93
$S + 2e^- \rightleftharpoons S^{2-}$	-0.48
$2SO_2(aq) + 2H^+ + 4e^- \rightleftharpoons S_2O_3^{2-} + H_2O$	-0.40
$SO_2(aq) + 4H^+ + 4e^- \rightleftharpoons S + 2H_2O$	$+0.50$
$S_4O_6^{2-} + 2e^- \rightleftharpoons 2S_2O_3^{2-}$	$+0.08$
$SO_4^{2-} + H_2O + 2e^- \rightleftharpoons SO_3^{2-} + 2OH^-$	-0.94
$2SO_4^{2-} + 4H^+ + 2e^- \rightleftharpoons S_2O_6^{2-} + 2H_2O$	-0.25
$S_2O_8^{2-} + 2e^- \rightleftharpoons 2SO_4^{2-}$	$+1.96$
$Sn^{2+} + 2e^- \rightleftharpoons Sn$	-0.14
$Sn^{4+} + 2e^- \rightleftharpoons Sn^{2+}$	$+0.15$
$Sr^{2+} + 2e^- \rightleftharpoons Sr$	-2.89
$Ti^{2+} + 2e^- \rightleftharpoons Ti$	-1.63
$Ti^{3+} + e^- \rightleftharpoons Ti^{2+}$	-1.37
$TiO^{2+} + e^- \rightleftharpoons Ti^{3+}$	$+0.10$
$Tl^+ + e^- \rightleftharpoons Tl$	-0.34
$V^{2+} + 2e^- \rightleftharpoons V$	-1.13
$V^{3+} + e^- \rightleftharpoons V^{2+}$	-0.26

E°/V

$$VO_2^+ + 2H^+ + 2e^- \rightleftharpoons V^{3+} + H_2O \qquad +0.34$$
$$VO_2^+ + 2H^+ + e^- \rightleftharpoons VO^{2+} + H_2O \qquad +1.00$$
$$Zn^{2+} + 2e^- \rightleftharpoons Zn \qquad -0.76$$
$$Zn(OH)_4^{2-} + 2e^- \rightleftharpoons Zn + 4OH^- \qquad -1.29$$

Index

a.c. voltammetry 59–60
activated complex 20
activity 9
activity coefficient 10
adsorptive stripping voltammetry
   66–7
amperometric titration 35
amperometry, definition of 3
anodic stripping voltammetry 64–5

batch injection analysis (BIA) 30
biamperometric titrations 35
bioelectroanalysis 74
biosensors 45, 61, 74
bipotentiometric titrations 35
Butler–Volmer equation 19

calomel electrode 11
capillary electrophoresis with
   electrochemical detection 73
carbon electrodes 13
cathodic stripping voltammetry 65
cell potential 10
channel electrode, double, cell with 29
   flow profile for 29
chemically modified electrodes 61–2
chronoamperogram, potential step 18
chronoamperometry, definition of 17
   potential step 17
chronopotentiogram, current step 19
chronopotentiometry, definition of 19
   current step 19
Clark oxygen electrode 61
coated wire electrodes 45
collection efficiency 31–3
   steady-state 32
conductimetry, definition of 2
conducting polymers 62, 71
continuous flow analysis 72–3
controlled current methods, see
   chronopotentiometry
convection 16, 26
Cottrell equation 17
current potential curve, irreversible
   reaction 24
   reversible reaction 23

cyclic voltammetry, adsorbed
   species 52
   experimental basis 49–50
   irreversible system 50–1
   microelectrodes 54
   multi-step reactions 53
   quasi-reversible system 52
   reversible system 50–1

data analysis 76–82
   detection limits from calibration
      plots 80–1
   F-test 79–80
   factorial analysis 81
   mean 76
   Q-test 77–8
   standard deviation 77
   t-test 78–9
diagnostic criteria for reversible
   reactions 23
   in cyclic and linear sweep
      voltammetry 51
differential pulse voltammetry 56–7
diffusion 16
   Fick's laws of 17
diffusion coefficient 17
diffusion layer 18
Donnan potential 15
double hydrodynamic electrodes 31–3
dropping mercury electrode (DME) 31
   diffusion limited current at 28

electrode auxiliary 13
   enzyme-selective 44–5
   gas-sensing 43–4
   glass 39–42
   indicator 12
   membrane-covered 61
   mercury thin film 64
   metallized membrane 61
   reference 11–12
   working 12
electrode kinetics, Butler Volmer
   formulation of 19–20
electrode materials 12–13
electrode-solution interface 25–6

electrolyte, supporting 12, 16
electrolytic cell 9
electron transfer: mechanism of 15–16
exchange current 20

Fermi level 8
Fick's laws 17
flow analysis, potentiometric
   sensors 47
   voltammetric sensors 67–8
flow injection analysis (FIA) 30
formal potential 9–10

galvanic cell 9
galvanostat 13
gas-sensing electrode 43–4
glass electrode 39–42
Gran plot 35

half-cell 10
half-peak potential in reversible
   LSV 51
half-wave potential 23
hanging mercury drop electrode 31
HPLC with electrochemical
   detection 67, 73
hydrodynamic electrodes 26–33
   coordinates for 27
   double 31–33
      collection efficiency 31–2
      shielding factor 32
   limiting currents at 28
hyphenated techniques 73–4

Ilkovic equation 28
injection analysis 72–3
   batch injection analysis 30, 72
   flow injection analysis 30, 72
inner Helmholtz plane 25
ion-selective electrodes, basis of
   functioning 38–9
   crystalline membrane 42–3
      heterogeneous 43
      homogeneous 42

detection limit, definition of 46
glass 39–42
non-crystalline membrane 43
ion exchange 43
neutral carrier 43
irreversible reaction 24
ISFETs 45–6

kinetics, electrode reaction 19–20

Lewis-Sargent relation 15
limiting currents at hydrodynamic
electrodes, table of 28
linear sweep voltammetry,
experimental basis 49–50
irreversible system 51
reversible system 50–1
liquid junction potential 14–15

mass transfer coefficient 16
mass transport, types of 16
mediators 62
membrane potential 15
mercury thin-film electrode 64
metallized membrane electrode 61
microelectrodes 33–4
diffusion-limited current 33
in voltammetric analysis 68, 71–2
migration 16
modified electrodes 61–2
voltammetric sensors 70–1

Nernst equation 9
Nicolsky–Eisenman equation 39

normal pulse voltammetry 55–6

outer Helmholtz plane 25
oxygen, removal of 4

peak current in cyclic voltammetry,
irreversible reaction 51
reversible reaction 51
polymer modified electrodes 62, 71
potential step, diffusion-limited current
due to 17
potential sweep, *see* linear sweep
voltammetry
potentiometric selectivity
coefficient 38–9
potentiometric sensors, novel 69–70
potentiometric stripping analysis 66
potentiometric titrations 34–5
potentiometry, definition of 2
potentiostat 13
preconcentration techniques 62–7
pulse voltammetry 54–9
differential pulse 56–7
normal pulse 55–6
square wave 58–9

quasi-reversible reactions, cyclic
voltammetry 52–3

rate constant, Butler–Volmer
expression for 19
RDE, *see* rotating disc electrode
reference electrodes 11–12
reversible reaction 22–3
cyclic voltammetry 51

rotating disc electrode, limiting current
at 28
schematic streamlines 28

salt bridge 10
sampled-current voltammetry, *see*
pulse voltammetry
Sand equation 19
self-assembled monolayers
(SAMs) 70–1
sonotrodes 30
square wave voltammetry 58–9
standard electrode potentials, table
of 83–6
standard hydrogen electrode 9
standard rate constant 19
static mercury drop electrode 31
stripping voltammetry 62–7

Tafel law 24–5
corrected for mass transport 25
titration, amperometric 35–6
biamperometric 35
bipotentiometric 35
potentiometric 34–5
transfer coefficient 20
tubular electrode, double (TDE) 29
flow profile at 29

voltammetry, definition of 2

wall-jet electrode, cell for 29
schematic streamlines 29